An Excursion Through Discrete Differential Geometry

Proceedings of Symposia in
APPLIED MATHEMATICS

Volume 76

An Excursion Through Discrete Differential Geometry

AMS Short Course
Discrete Differential Geometry
January 8–9, 2018
San Diego, California

Keenan Crane
Editor

AMERICAN MATHEMATICAL SOCIETY
Providence, Rhode Island

LECTURE NOTES PREPARED FOR THE AMERICAN MATHEMATICAL SOCIETY SHORT COURSE ON DISCRETE DIFFERENTIAL GEOMETRY HELD IN SAN DIEGO, CALIFORNIA, JANUARY 8–9, 2018

The AMS Short Course Series is sponsored by the Society's Program Committee for National Meetings. The series is under the direction of the Short Course Subcommittee of the Program Committee for National Meetings.

2010 *Mathematics Subject Classification.* Primary 53-xx, 52Cxx, 65-xx.

Library of Congress Cataloging-in-Publication Data

Names: American Mathematical Society. Short Course, Discrete Differential Geometry (2018: San Diego, Calif.), author. | Crane, Keenan, 1984– editor.
Title: An excursion through discrete differential geometry: AMS Short Course, Discrete Differential Geometry, January 8–9, 2018, San Diego, California / Keenan Crane, editor.
Description: Providence, Rhode Island: American Mathematical Society, [2020] | Series: Proceedings of symposia in applied mathematics, 0160-7634; volume 76 | "Lecture notes prepared for the American Mathematical Society course Discrete Differential Geometry held in San Diego, California, January 8-9, 2018." | Includes bibliographical references.
Identifiers: LCCN 2020014948 | ISBN 9781470446628 (softcover) | ISBN 9781470460037 (ebook)
Subjects: LCSH: Geometry, Differential–Congresses. | Discrete geometry–Congresses. | Numerical analysis–Congresses. | AMS: Differential geometry. | Convex and discrete geometry – Discrete geometry. | Numerical analysis.
Classification: LCC QA649 .A482 2020 | DDC 516.3/6–dc23
LC record available at https://lccn.loc.gov/2020014948
DOI: https://doi.org/10.1090/psapm/076

∞ The paper used in this book is acid-free and falls within the guidelines
established to ensure permanence and durability.
Visit the AMS home page at `https://www.ams.org/`

10 9 8 7 6 5 4 3 2 1   25 24 23 22 21 20

Contents

Preface

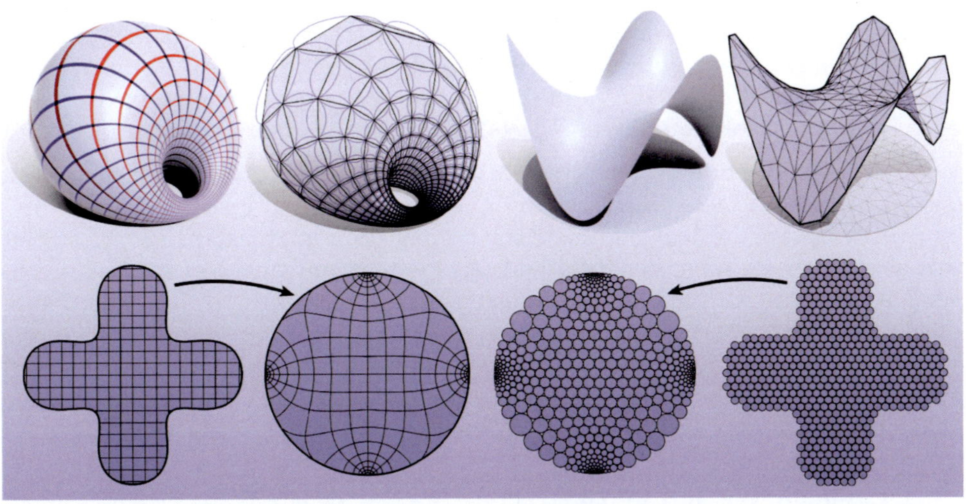

FIGURE 1. Discrete differential geometry re-imagines classical ideas from differential geometry without reference to differential calculus. For instance, surfaces parameterized by principal curvature lines are replaced by meshes made of circular quadrilaterals (top left), the maximum principle obeyed by harmonic functions is expressed via conditions on the geometry of a triangulation (top right), and complex-analytic functions can be replaced by so-called *circle packings* that preserve tangency relationships (bottom). These discrete surrogates provide a bridge between geometry and computation, while at the same time preserving important structural properties and theorems.

1. Overview

The emerging field of discrete differential geometry (DDG) studies discrete analogues of smooth geometric objects, providing an essential link between analytical descriptions and computation[1] (see Figure 1). In recent years it has unearthed a rich variety of new perspectives on applied problems in computational anatomy/biology, computational mechanics, industrial design, computational architecture, and digital geometry processing at large.

The basic philosophy of discrete differential geometry is that a discrete object like a polyhedron is not merely an approximation of a smooth one, but rather a differential geometric object in its own right. In contrast to traditional numerical analysis, which focuses on eliminating approximation error in the limit of refinement (e.g., by taking smaller and smaller finite differences), DDG places an emphasis on the so-called "mimetic" viewpoint, where key properties of a system are preserved exactly, independent of how large or small the elements of a mesh might be. Just as algorithms for simulating mechanical systems might seek to exactly preserve physical invariants such as total energy or momentum, structure-preserving models of discrete geometry seek to exactly preserve global geometric invariants such as total curvature. More broadly, DDG focuses on the discretization of objects that do not naturally fall under the umbrella of traditional numerical analysis.

The Game. The spirit of discrete differential geometry is well-illustrated by a "game" often used to develop discrete analogs of a given smooth object:

(1) Write down several equivalent definitions in the smooth setting.
(2) Apply each smooth definition to an object in the discrete setting.
(3) See which properties of the original smooth object are preserved by each of the resulting discrete objects, which are invariably inequivalent.

Most often, no discrete object preserves all the properties of the original smooth one—a so-called *no free lunch* scenario. Nonetheless, the properties that are preserved often prove invaluable for particular applications and algorithms. Moreover, this activity yields some beautiful and unexpected consequences, such as a connection between conformal geometry and pure combinatorics, or a description of constant-curvature surfaces that requires no definition of curvature! These notes provide an incomplete overview of several contemporary topics in DDG; a broad overview can also be found in the recent *Notices* article, "A Glimpse into Discrete Differential Geometry" by Crane and Wardetzky.

1.1. Geometry and finitism. There is a long debate in mathematics about the meaning and implications of various notions of infinity dating back at least to Aristotle and brought to a head in the late 19th century by Cantor. *Finitists* take the view that (in one way or another) the only "real" objects are those that have finite descriptions (perhaps allowing for "potential infinities," like the natural numbers). For some, limiting mathematics to discrete or finite objects is a deep philosophical conviction; for others, it is an entertaining game. In the context of computation, the search for finite descriptions has a more pragmatic motivation: Turing's abstract model of computation is discrete; real physical computing devices

[1]This preface is adapted from Crane and Wardetzky, "A Glimpse into Discrete Differential Geometry," *Notices of the American Mathematical Society*, Vol. 64, No. 10, pp. 1153–1159, 2017.

are necessarily finite. If one wishes to use machines to analyze geometry, one must therefore have good finite models[2].

Since one knows a priori that geometry must have a finite description, it seems worthwhile to consider geometric models that directly represent the way geometry behaves in nature, rather those that converge to the true behavior only in an unattainable limit. In fact, this situation leads to a natural question: Is some notion of infinity strictly required to faithfully model nature? Or can the basic things we want to say about natural geometry be captured by purely discrete models? The differential calculus of Leibniz and Newton has proven to be an unreasonably effective tool for modeling the natural world. For this reason it is tempting to feel that models based on differential equations are the true, canonical Platonic forms, and anything discrete and finite is a mere approximation. Yet nobody denies that even beloved continuum models (say, the Navier-Stokes equations) break down at very small scales, where nature looks increasingly like quantized packets of matter and energy. From this point of view, one might also be justified in feeling like nature is inherently discrete and that continuum models are perceived as "correct" only because they closely approximate observed behavior at the macroscopic scale. Moreover, infinity can creep into geometry in ways that betray our experience of nature. For instance, the *Banach-Tarski paradox* allows (by sly use of infinity) a solid ball to be decomposed into finitely many pieces, then reassembled into two balls—each of the same volume as the original one!

The purpose of mentioning such examples is not to ridicule the use of infinity, but rather to say: in order for a discrete model to be a "first class" description of an object, it need not duplicate all the strange behavior found in the continuous setting. One should instead ask, on both the smooth and discrete side: which phenomena are fundamental, and which are purely artifacts of the language in which the model is expressed? In fact, the dual perspectives offered by smooth and discrete differential geometry help to distill which phenomena are fundamental and which are superficial, since the most fundamental properties tend to arise naturally and easily on both sides (say, conservation laws in classical mechanics or global topological invariants in Riemannian geometry). In other words, geometry is neither inherently smooth nor discrete; rather, it is comprised of all the ideas that persist independent of the particular language used to write them down.

One can also tell this story from an analytical point of view: by considering only finite, discrete models, one obtains a degree of regularity that naturally excludes certain strange examples and behaviors. A nice example is the question of whether a 2-sphere embedded in $\mathbb{R}^3$ always divides space into two simply-connected regions. Intuitively, the answer is "yes," since the 2-sphere is simply connected, and embeddings preclude any change in global topology. Yet one can construct a topological embedding (again by sly use of infinity) called *Alexander's horned sphere*, where a loop around one of the "horns" can never be contracted. On the other hand, Alexander himself showed that if one restricts the question to discrete, or more precisely, *piecewise linear* embeddings, then no such example is possible. In other words, working in the simplicial category yields a result that is perhaps more in line with the kind of behavior that we would expect to encounter in nature. (Likewise, Banach-Tarski has no analogue in the finite, simplicial setting.)

[2]Note that even the symbolic expressions used to reason about geometry in a computer algebra system—or with pen and paper—are finite expressions in a formal grammar!

A more negative point of view—encapsulated by the *no free lunch* situation—is that discrete models are ultimately too rigid to describe all the geometric behavior observed in nature. The question of rigidity is a central and ongoing question in discrete differential geometry and a major theme of these notes. On the one hand, it is often true that no single discrete object in a certain class can faithfully reproduce a given list of properties from the smooth setting (see for instance our discussion of discrete Laplace operators). Are such results the end of the story, or does one simply need to consider an alternative approach to discretization? By adopting a different point of view, as outlined in The Game, situations that initially look hopeless sometimes give way to rich and flexible theories that capture much of the structure found in the smooth setting (such as conformal equivalence of triangle meshes). A major goal of discrete differential geometry is to see how far we can push the "finitist" point of view, not only to enable practical computation, but also to obtain clearer descriptions and a deeper understanding of the way shape behaves in nature.

Keenan Crane

Proceedings of Symposia in Applied Mathematics
Volume **76**, 2020
https://doi.org/10.1090/psapm/076/00669

Discrete Laplace operators

Max Wardetzky

In this chapter[1] we review some important properties of Laplacians, smooth and discrete. We place special emphasis on a unified framework for treating smooth Laplacians on Riemannian manifolds alongside discrete Laplacians on graphs and simplicial manifolds. We cast this framework into the language of linear algebra, with the intent to make this topic as accessible as possible. We combine perspectives from smooth geometry, discrete geometry, spectral analysis, machine learning, numerical analysis, and geometry processing within this unified framework.

1. Introduction

The Laplacian is perhaps *the* prototypical differential operator for various physical phenomena. It describes, for example, heat diffusion, wave propagation, steady state fluid flow, and it is key to the Schrödinger equation in quantum mechanics. In Euclidean space, the Laplacian of a smooth function $u : \mathbb{R}^n \to \mathbb{R}$ is given as the sum of second partial derivatives along the coordinate axes,

$$\Delta u = -\left(\frac{\partial^2 u}{\partial x_1^2} + \frac{\partial^2 u}{\partial x_2^2} + \cdots + \frac{\partial^2 u}{\partial x_n^2}\right),$$

where we adopt the geometric perspective of using a minus sign.

1.1. Basic properties of Laplacians. The Laplacian has many intriguing properties. For the remainder of this exposition, consider an open and bounded domain $\Omega \subset \mathbb{R}^n$ and the L^2 inner product

$$(f, g) := \int_\Omega fg$$

on the linear space of square-integrable functions on Ω. Let $u, v : \Omega \to \mathbb{R}$ be two (sufficiently smooth) functions that vanish on the boundary of Ω. Then the Laplacian Δ is a *symmetric* (or, to be precise, a formally self-adjoint) linear operator with respect to this inner product since integration by parts yields

$$\text{(Sym)} \qquad (u, \Delta v) = \int_\Omega \nabla u \cdot \nabla v = (\Delta u, v).$$

2020 *Mathematics Subject Classification.* Primary 65M99; Secondary 65N99.
Key words and phrases. Discrete differential geometry, Laplace-Beltrami.

[1]This chapter is republished with permission from Taylor & Francis Group LLC - Books, from "Generalized Barycentric Coordinates in Computer Graphics and Computational Mechanics", Kai Hormann and N. Sukumar, 2017. Permission conveyed through Copyright Clearance Center, Inc.

Here ∇ denotes the standard gradient operator and $\nabla u \cdot \nabla v$ denotes the standard inner product between vectors in $\mathbb{R}^n$. The choice of using a minus sign in the definition of the Laplacian makes this operator positive semi-definite since

$$(u, \Delta u) = \int_\Omega \nabla u \cdot \nabla u \geq 0. \tag{PSD}$$

If one restricts to functions that vanish on the boundary of Ω, PSD implies that the only functions that lie in the kernel of the Laplacian ($\Delta u = 0$) are those functions that vanish on the entire domain. Moreover, properties SYM and PSD imply that the Laplacian can be diagonalized and its eigenvalues are nonnegative,

$$\Delta u = \lambda u \Rightarrow \lambda \geq 0.$$

Another prominent property of smooth Laplacians is the *maximum principle.* Let $u : \Omega \to \mathbb{R}$ be *harmonic*, i.e., $\Delta u = 0$. The maximum principle asserts that

$$u \text{ is harmonic} \Rightarrow u \text{ has no strict local maximum in } \Omega, \tag{MAX}$$

where we no longer assume that u vanishes on the boundary of Ω. Likewise, no harmonic function can have a strict local minimum in Ω.

The maximum principle can be derived from another important property of harmonic functions, the *mean value property.* Consider a point $\mathbf{x} \in \Omega$ and a closed ball $B(\mathbf{x}, r)$ of radius r centered at $\mathbf{x}$ that is entirely contained in Ω. Every harmonic function has the property that the value $u(\mathbf{x})$ can be recovered from the average of the values of u in the ball $B(\mathbf{x}, r)$:

$$u(\mathbf{x}) = \frac{1}{\operatorname{vol}(B(\mathbf{x}, r))} \int_{B(\mathbf{x},r)} u(\mathbf{y}) d\mathbf{y}.$$

A simple argument by contradiction shows that the mean value property implies property MAX.

The properties mentioned so far play an important role in applications; specifically, in the context of barycentric coordinates, they give rise to *harmonic coordinates* and *mean value coordinates*, see [**Flo03, JMD⁺07, JSW05**]. Below we discuss additional properties of Laplacians. For further reading we refer to the books [**Ber03, Eva98, Ros97**] and the lecture notes [**Can13, CdGDS13**].

2. Smooth Laplacians on Riemannian manifolds

The standard Laplacian in $\mathbb{R}^n$ can be expressed as

$$\Delta u = -\operatorname{div} \nabla u,$$

where div is the usual divergence operator acting on vector fields in $\mathbb{R}^n$. Written in integral form, the negative divergence operator is the (formal) adjoint of the gradient: If X is a vector field and $u : \Omega \to \mathbb{R}$ is a function that vanishes on the boundary of Ω, then

$$\int_\Omega \nabla u \cdot X = \int_\Omega u\,(-\operatorname{div} X).$$

This perspective can be generalized to Riemannian manifolds. The Laplacian plays an important role in the study of these curved spaces.

2.1. Exterior calculus. Although gradient and divergence can readily be defined on Riemannian manifolds, it is more convenient to work with the differential (or *exterior derivative*) d instead of the gradient and with the codifferential d^* instead of divergence.

The differential d is similar to (but not the same as) the gradient. Indeed, given a function $u : \Omega \to \mathbb{R}$, one has

$$du(X) = \nabla u \cdot X$$

for every vector field X. In particular, the differential does not require the notion of a metric, whereas the gradient does. The codifferential d^* is defined as the formal adjoint (informally, transpose) to d, in the same way as divergence is the adjoint of the gradient. In contrast to the divergence operator, which acts on vector fields, the codifferential d^* acts on 1-forms. A 1-form is a covector at every point of Ω, i.e., if X is a vector field on Ω and α is a 1-form, then $\alpha(X)$ is a real-valued function on Ω. In order to define the codifferential d^*, consider a 1-form α and a function $u : \Omega \to \mathbb{R}$ that vanishes on the boundary of Ω. Then

$$\int_\Omega du \cdot \alpha = \int_\Omega u\, d^* \alpha,$$

where the dot product is the inner product between covectors induced from the inner product between vectors. Notice that different from the differential d, the codifferential *does* require the notion of a metric. The Laplacian of a function u can be expressed as

$$\Delta u = d^* du,$$

which is equivalent to the representation $\Delta u = -\text{div}\nabla u$ given above.

In order to carry over this framework to manifolds, let M be a smooth orientable manifold with smooth Riemannian metric g. Suppose for simplicity that M is compact and has empty boundary. The Riemannian metric induces a pointwise inner product between tangent vectors on M, which, analogously to the above discussion, induces an inner product between 1-forms. More generally, one works with k-forms for $k \geq 0$. A 0-form, by convention, is a real-valued function on M. A 1-form can be thought of as an oriented 1-volume in the sense that applying a 1-form to a vector field returns a real value at every point. Likewise, a k-form for $k > 1$ can be thought of as an oriented k-volume in the sense of returning a real number at every point when applied to an ordered k-tuple (parallelepiped) of tangent vectors. As a consequence, k-forms can be integrated over (sub)manifolds of dimension k. In the sequel, we let Λ^k denote the linear space of k-forms on M.

Analogous to the L^2 inner product between function in $\mathbb{R}^n$, let

$$(\alpha, \beta)_k := \int_M g(\alpha, \beta)\text{vol}_g$$

denote the L^2 inner product between k-forms α and β on M, where, by slight abuse of notation, we let $g(\alpha, \beta)$ denote the (pointwise) inner product induced by the Riemannian metric.

The differential $d : \Lambda^k \to \Lambda^{k+1}$ maps k-forms to $(k+1)$-forms for $0 \leq k \leq \text{dim}M$, where one sets $d\alpha = 0$ for any k-form with $k = \text{dim}M$. One can define the differential acting on k-forms by postulating *Stokes' theorem*,

$$\int_U d\alpha = \int_{\partial U} \alpha,$$

for every k-form α and every (sufficiently smooth) submanifold $U \subset M$ of dimension $(k+1)$ with boundary ∂U. If one asserts this equality as the defining property of the differential d, then it immediately follows that $d \circ d = 0$ since the boundary of a boundary of a manifold is empty ($\partial(\partial U) = \emptyset$).

The codifferential d^*, taking $(k+1)$-forms back to k-forms, is the (formal) adjoint of d with respect to the L^2 inner products on k- and $(k+1)$-forms. It is defined by requiring that

$$(d\alpha, \beta)_{k+1} = (\alpha, d^*\beta)_k$$

for all k-forms α and all $(k+1)$-forms β. Finally, the *Laplace–Beltrami operator* $\Delta : \Lambda^k \to \Lambda^k$ acting on k-forms is defined as

$$\Delta\alpha := dd^*\alpha + d^*d\alpha.$$

Notice that this expression reduces to $\Delta u = d^*du$ for 0-forms (functions) on M. It follows almost immediately from the definition of the Laplacian that a k-form α is *harmonic* ($\Delta\alpha = 0$) if and only if α is closed ($d\alpha = 0$) and co-closed ($d^*\alpha = 0$).

From a structural perspective it is important to note that properties SYM, PSD, and MAX mentioned earlier remain true (among various other properties) in the setting of Riemannian manifolds. For further details on exterior calculus and the Laplace–Beltrami operator, we refer to [**Ros97**].

2.2. Hodge decomposition. Every sufficiently smooth k-form α on M admits a unique decomposition

$$\alpha = d\mu + d^*\nu + h,$$

known as the *Hodge decomposition* (or Hodge–Helmholtz decomposition), where μ is a $(k-1)$-from, ν is a $(k+1)$-form and h is a harmonic k-form. This decomposition is unique and orthogonal with respect to the L^2 inner product on k-forms,

$$0 = (d\mu, d^*\nu)_k = (h, d\mu)_k = (h, d^*\nu)_k,$$

which immediately follows from the fact that $d \circ d = 0$ and the fact that harmonic forms satisfy $dh = d^*h = 0$. The Hodge decomposition can be thought of as a (formal) application of the well-known fact from linear algebra that the orthogonal complement of the kernel of a linear operator is equal to the range of its adjoint (transpose) operator.

By duality between vector fields and 1-forms, the Hodge decomposition for 1-forms carries over to a corresponding decomposition for vector fields into curl-free and divergence-free components, which has applications for fluid mechanics [**AK98**] and Maxwell's equations for electromagnetism [**Fra04**].

Geometrically, the Hodge decomposition establishes relations between the Laplacian and global properties of manifolds. Indeed, the linear space of harmonic k-forms is finite-dimensional for compact manifolds and isomorphic to $H^k(M;\mathbb{R})$, the k-th cohomology of M. As an application of this fact, consider a compact orientable surface without boundary. Then the dimension of the space of harmonic 1-forms is equal to twice the genus of the surface, that is, this dimension is zero for the 2-sphere, two for the two-dimensional torus, four for a genus two surface (pretzel) and so on. Hence the Laplacian provides global information about the topology of the underlying space.

For a thorough treatment of Hodge decompositions, including the case of manifolds with boundary, we refer to [**Sch95**].

2.3. The spectrum. One cannot speak about the Laplacian without discussing its spectrum. On a compact orientable manifold without boundary, it follows from the inequality

$$(\Delta u, u)_0 = (du, du)_1 \geq 0$$

that the spectrum is nonnegative and that the only functions in the kernel of the Laplacian are constant functions. Thus zero is a trivial eigenvalue of the Laplacian with a one-dimensional space of eigenfunctions. The next (non-trivial) eigenvalue $\lambda_1 > 0$ is much more interesting. By the min-max principle, this eigenvalue satisfies

$$\lambda_1 = \min_{(u,1)_0=0} \frac{(du, du)_1}{(u, u)_0},$$

where one takes the minimum over all functions that are L^2-orthogonal to the constants. Higher eigenvalues can be obtained by successively applying the min-max principle to the orthogonal complements of the eigenspaces of lower eigenvalues.

The first non-trivial eigenvalue λ_1 tells a great deal about the geometry of the underlying Riemannian manifold. As an example, consider Cheeger's isoperimetric constant

$$\lambda_C := \inf_{N \neq \emptyset} \left\{ \frac{\mathrm{vol}_{n-1}(N)}{\min(\mathrm{vol}_n(M_1),\, \mathrm{vol}_n(M_2))} \right\},$$

where N runs over all compact codimension-1 submanifolds that partition M into two disjoint open sets M_1 and M_2 with $N = \partial M_1 = \partial M_2$. Intuitively, the optimal N for which λ_C is attained partitions M into two sets that have maximal volume and minimal perimeter. As an example, suppose that M has the shape of the surface of a smooth dumbbell. Then N is a curve going around the axis of the dumbbell at the location where the dumbbell is thinnest.

A relation of Cheeger's constant to the first non-trivial eigenvalue of the Laplacian is provided by the Cheeger inequalities

$$\frac{\lambda_C^2}{4} \leq \lambda_1 \leq c(K\lambda_C + \lambda_C^2),$$

where the constant c only depends on dimension and $K \geq 0$ provides a lower bound on the Ricci curvature of M in the sense that $Ric(M, g) \geq -K^2(n-1)$; see [**Bus82**, **Cha84**]. Recall that for surfaces, Ricci curvature and Gauß curvature coincide. The first non-trivial eigenvalue of the Laplacian is thus related to the metric problem of minimal cuts—thus providing a relation between an analytical quantity (the first eigenvalue) and a purely geometric quantity (the Cheeger constant). Intuitively, if λ_1 is small, then M must have a small bottleneck; vice-versa, if λ_1 is large, then M is somewhat thick.

Equipped with the full set of eigenfunctions $\{\varphi_i\}$ of the Laplace–Beltrami operator, one can perform Fourier analysis on manifolds by decomposing any square-integrable function u into its Fourier-modes,

$$u = \sum_i (u, \varphi_i)_0 \varphi_i,$$

provided that one chooses the eigenfunctions such that $(\varphi_i, \varphi_i) = \delta_{ij}$. (Notice that $(\varphi_i, \varphi_j)_0 = 0$ is automatic for eigenfunctions belonging to different eigenvalues.) The Fourier perspective is of great relevance in signal and geometry processing.

Maintaining a spectral eye on geometry, it is natural to ask the inverse question: How much *geometric* information can be reconstructed from information about the

Laplacian? If the entire Laplacian is known on a smooth manifold, then one can reconstruct the metric, for example, by using the expression of Δ in local coordinates. If, however, "only" the spectrum is known, then less can be said in general. For example, Kac's famous question *Can one hear the shape of a drum?* [**Kac66**], that is, whether the entire geometry can be inferred from the spectrum alone, has a negative answer: There exist isospectral but non-isometric manifolds [**GWW92**, **Sun85**].

3. Discrete Laplacians

Discrete Laplacians can be defined on simplicial manifolds or, more generally, on graphs. We treat the case of graphs first and discuss simplicial manifolds further below. We let our discussion of discrete Laplacians be guided by drawing upon the smooth setup above.

3.1. Laplacians on graphs. Consider an undirected graph $\Gamma = (V, E)$ with vertex set V and edge set E. For simplicity we only consider finite graphs here. Suppose that every edge $e \in E$ between vertices $i \in V$ and $j \in V$ carries a real-valued weight $\omega_e = \omega_{ij} = \omega_{ji} \in \mathbb{R}$. We discuss below how weights can be chosen; suppose for now that such a choice has been made. A discrete Laplacian acting on a function $u : V \to \mathbb{R}$ is defined as

$$(Lu)_i := \sum_{j \sim i} \omega_{ij}(u_i - u_j), \tag{3.1}$$

where the sum ranges over all vertices j that are connected by an edge with vertex i.[2] This allows for representing the linear operator L as a matrix by

$$L_{ij} := \begin{cases} -\omega_{ij} & \text{if there is an edge between } i \text{ and } j, \\ \sum_{k \sim i} \omega_{ik} & \text{if } i = j, \\ 0 & \text{otherwise.} \end{cases}$$

The matrix L is called the discrete Laplace matrix. This definition may seem to come a bit out of the blue. In order to see how it relates to smooth Laplacians, consider again the smooth case and the quantity

$$E_D[u] := \frac{1}{2} \int_\Omega \|\nabla u\|^2,$$

which is known as the *Dirichlet energy* of u. In the discrete setup, one may discretize the gradient ∇u along an edge $e = (i, j)$ as the finite difference $(u_i - u_j)$. Accordingly, one defines discrete Dirichlet energy as

$$E_D[u] := \frac{1}{2} \sum_{e \in E} \omega_{ij}(u_i - u_j)^2,$$

where the sum ranges over all edges. One then has

$$\text{(discrete)}\ E_D[u] = \frac{1}{2} u^T L u \quad \text{vs.} \quad \text{(smooth)}\ E_D[u] = \frac{1}{2}(u, \Delta u)_0,$$

which justifies calling L a Laplace matrix. Due to this representation (and using the language of partial differential equations [**Eva98**]) we call discrete Laplacians of the

---

[2] Some authors include a division by vertex weights in the definition of Laplacians on graphs. Such a division arises naturally when considering strongly defined Laplacian, instead of weakly defined Laplacians. We come back to this distinction below.

form (3.1) *weakly defined* instead of *strongly defined.* We return to this distinction below.

Weakly defined discrete Laplacians of the form (3.1) are always symmetric—they satisfy SYM due to the assumption that $\omega_{ij} = \omega_{ji}$. However, whether or not a discrete Laplacian satisfies (at least some of) the other properties of the smooth setting heavily depends on the choice of weights.

The simplest choice of weights is to set $\omega_{ij} = 1$ whenever there is an edge between vertices i and j. This results in the so-called *graph Laplacian.* The diagonal entries of the graph Laplacian are equal to the degree of the respective vertex, that is, the number of edges adjacent to that vertex. The graph Laplacian is just a special case of what we call a *Laplacian on graphs* here.

Positive edge weights are a natural choice if weights resemble transition probabilities of a random walker. Discrete Laplacians with positive weights are always positive semi-definite PSD and, just like in the smooth setting, they only have the constant functions in their kernel provided that the graph is connected. As a word of caution we remark that positivity of weights is *not* necessary to guarantee PSD. Below we discuss Laplacians that allow for (some) negative edge weights but still satisfy PSD.

Laplacians with positive edge weights always satisfy the mean value property since every harmonic function u (a function for which $Lu = 0$) satisfies

$$u_i = \sum_{j \sim i} l_{ij} u_j \quad \text{with} \quad l_{ij} = \frac{\omega_{ij}}{L_{ii}} > 0.$$

Therefore, discrete Laplacians with positive weights also satisfy the maximum principle MAX since $\sum_{j \sim i} l_{ij} = 1$ and thus u_i is a convex combination of its neighbors u_j for discrete harmonic functions.

3.2. The spectrum. As in the smooth case, one cannot discuss discrete Laplacians without mentioning their spectrum and their eigenfunctions, which provide a fingerprint of the structure of the underlying graph.

As an example consider again Cheeger's isoperimetric constant. In order to define this constant in the discrete setting consider a partitioning of Γ into two disjoint subgraphs Γ_1 and $\bar{\Gamma}_1$ such that the vertex set V of Γ is the disjoint union of the vertex sets V_1 of Γ_1 and $\bar{V}_1$ of $\bar{\Gamma}_1$. Here a subgraph of Γ denotes a graph whose vertex set is a subset of the vertex set of Γ such that two vertices in the subgraph are connected by an edge if and only if they are connected by an edge in Γ. For positive edge weights, the discrete Cheeger constant (sometimes called *conductance* of a weighted graph) is defined as

$$\lambda_C := \min \left\{ \frac{\operatorname{vol}(\Gamma_1, \bar{\Gamma}_1)}{\min(\operatorname{vol}(\Gamma_1), \operatorname{vol}(\bar{\Gamma}_1))} \right\},$$

where

$$\operatorname{vol}(\Gamma_1, \bar{\Gamma}_1) := \sum_{i \in V_1, j \notin V_1} \omega_{ij} \quad \text{and} \quad \operatorname{vol}(\Gamma_1) := \sum_{i \in V_1, j \in V_1} \omega_{ij},$$

and similarly for $\operatorname{vol}(\bar{\Gamma}_1)$. Notice that for the case of the graph Laplacian $\operatorname{vol}(\Gamma_1)$ equals twice the number of edges in Γ_1 and $\operatorname{vol}(\Gamma_1, \bar{\Gamma}_1)$ equals the number of edges with one vertex in Γ_1 and another vertex in its complement.[3]

[3]Some authors use a different version of the respective volumes in the definition of the Cheeger constant, resulting in different versions of the Cheeger inequalities; see [**Sun07**].

Similar to the smooth case, one then obtains the Cheeger inequalities

$$\frac{\lambda_C^2}{2} \leq \tilde{\lambda}_1 \leq 2\lambda_C,$$

where $\tilde{\lambda}_1$ is the first nontrivial eigenvalue of the rescaled Laplace matrix

$$\tilde{L} := SLS,$$

where S is a diagonal matrix with $S_{ii} = 1/\sqrt{\omega_{ii}}$. This rescaling is necessary since λ_C is invariant under a uniform rescaling of edge weights (and so is $\tilde{L}$), whereas L scales linearly with the edge weights. A proof of the discrete Cheeger inequalities for the case of the graph Laplacians ($\omega_{ij} = 1$) can be found in [**Chu07**]; the proof for arbitrary positive edge weights is nearly identical.

The Cheeger constant—and alongside the corresponding partitioning of Γ into the two disjoint subgraphs Γ_1 and $\bar{\Gamma}_1$—has applications in graph clustering, since the edges connecting Γ_1 and $\bar{\Gamma}_1$ tend to "cut" the graph along its bottleneck [**KVV04**].

Any discrete Laplacian (having positive edge weights or not) that satisfies SYM and PSD can be used for Fourier analysis on graphs. Indeed, let $\{\varphi_i\}$ be the eigenfunctions of L, chosen such that $\varphi_i^T \varphi_j = \delta_{ij}$. Then one has

$$u = \sum_i (u^T \varphi_i)\varphi_i,$$

just like in the smooth setting. This decomposition of discrete functions on graphs into their Fourier modes has a plethora of applications in geometry processing; see [**LZ10**] and references therein.

As in the smooth setting, it is natural to ask the inverse question of how much geometric information can be inferred from information about the Laplacian. Recall that in the the smooth case, knowing the full Laplacian allows for recovering the Riemannian metric. Similarly, in the discrete case it can be shown that for simplicial surfaces the knowledge of the cotan weights (that are introduced below) for discrete Laplacian allows for reconstructing edge lengths (i.e., the discrete metric) of the underlying mesh up to a global scale factor [**ZGLG12**]. If, however, only the spectrum of the Laplacian is known, then there exist isospectral but non-isomorphic graphs [**Bro97**]. In fact, for the case of the cotan Laplacian (see below), the exact same isospectral domains considered in [**GWW92**] that were originally proposed for showing that "One cannot hear the shape of a drum" for smooth Laplacians work in the discrete setup.

A curious fact concerning the connection between discrete Laplacians and the underlying geometry is *Rippa's theorem* [**Rip90**]: The Delaunay triangulation of a fixed point set in $\mathbb{R}^n$ minimizes the Dirichlet energy of any piecewise linear function over this point set. In [**CXGL10**], this result is taken a step further, where the authors show that the spectrum of the cotan Laplacian obtains its minimum on a Delaunay triangulation in the sense that the i-th eigenvalue of the cotan Laplacian of any triangulation of a fixed point set in the plane is bounded below by the i-th eigenvalue resulting from the cotan Laplacian associated with the Delaunay triangulation of the given point set.

3.3. Laplacians on simplicial manifolds. Recall that in the smooth case, Laplacians acting on k-forms take the form

$$\Delta = dd^* + d^*d.$$

In order to mimic this construction in the discrete setting, one requires a bit more structure than just an arbitrary graph. To this end, consider a simplicial manifold, such as a triangulated surface. We keep referring to this manifold as M. As in the smooth case, for simplicity, suppose that M is orientable and has no boundary.

Simplicial manifolds allow for a natural definition of discrete k-forms as duals of k-cells. Indeed, every 0-form is a function defined on vertices, a 1-form α is dual to edges, that is, $\alpha(e)$ is a real number for any oriented 1-cell (oriented edge), a 2-form is dual to oriented 2-cells, and so forth. In the sequel, let the linear space of discrete k-forms (better known as simplicial cochains) be denoted by C^k.

As in the smooth case, the discrete differential δ (better known as the coboundary operator) maps discrete k-forms to discrete $(k+1)$-forms, $\delta : C^k \to C^{k+1}$. Again, as in the smooth case, the discrete differential can be defined by *postulating* Stokes' formula: Let α be a discrete k-form. Then one requires that

$$\delta\alpha(\sigma) = \alpha(\partial\sigma)$$

for all $(k+1)$-cells σ, wehere ∂ denotes the simplicial boundary operator. The simplicial boundary operator, when applied to a vertex returns zero (since vertices do not have a boundary). When applied to an oriented edge, the boundary operator returns the difference between the edge's vertices. Likewise, ∂ applied to an oriented 2-cell σ returns the sum of oriented edges of σ (with the orientation induced by that of σ). Since the boundary of a boundary is empty ($\partial \circ \partial = 0$) one has $\delta \circ \delta = 0$, just like in the smooth case. Notice that the definition of δ does not require the notion of inner products.

In order to define the discrete codifferential δ^* one additionally requires an inner product $(\cdot, \cdot)_k$ on the linear space of k-forms for each k. Below we discuss the construction of such inner products. Given a fixed choice of inner products on discrete k-forms, the codifferential is defined by requiring that

$$(\delta\alpha, \beta)_{k+1} = (\alpha, \delta^*\beta)_k$$

for all k-forms α and all $(k+1)$-forms β, and the discrete *strongly defined* Laplacian acting on k-forms takes the form

$$\mathbb{L} := \delta\delta^* + \delta^*\delta. \tag{3.2}$$

This perspective is that of *discrete exterior calculus* (DEC) [**CdGDS13, DHLM05**], where—by slight abuse of notation—inner products are referred to as "discrete Hodge stars".

Strongly defined Laplacians are self-adjoint with respect to the inner products $(\cdot, \cdot)_k$ on discrete k-forms, since

$$(\mathbb{L}\alpha, \beta)_k = (\delta\alpha, \delta\beta)_{k+1} + (\delta^*\alpha, \delta^*\beta)_{k-1} = (\alpha, \mathbb{L}\beta)_k.$$

Moreover, strongly defined Laplacians are always positive semi-definite PSD, since

$$(\mathbb{L}\alpha, \alpha)_k = (\delta\alpha, \delta\alpha)_{k+1} + (\delta^*\alpha, \delta^*\alpha)_{k-1} \geq 0.$$

In particular, a discrete k-form is harmonic ($\mathbb{L}\alpha = 0$) if and only if $\delta\alpha = \delta^*\alpha = 0$, just like in the smooth setting.

3.4. Strongly and weakly defined Laplacians. Every strongly defined Laplacian as given by (3.2) has a weakly defined cousin L acting on discrete functions u. The weak version is obtained by requiring that at every vertex i the resulting function Lu is equal to

$$(Lu)_i := (\mathbb{L}u, 1_i)_0 = (\delta^* \delta u, 1_i)_0 = (\delta u, \delta 1_i)_1,$$

where 1_i is the indicator function of vertex i. In particular, let $\mathbb{M}_0$ and $\mathbb{M}_1$ be the symmetric positive definite matrices that encode the inner products between 0-forms and 1-forms, respectively,

$$(u, v)_0 = u^T \mathbb{M}_0 v \quad \text{and} \quad (\alpha, \beta)_1 = \alpha^T \mathbb{M}_1 \beta.$$

Then the weakly and strongly defined Laplacians satisfy, respectively,

$$L = \delta^T \mathbb{M}_1 \delta \quad \text{and} \quad \mathbb{L} = \mathbb{M}_0^{-1} L.$$

As an example, consider diagonal inner products on 0-forms and 1-forms,

$$(u, v)_0 = \sum_{i \in V} m_i u_i v_i \quad \text{and} \quad (\alpha, \beta)_1 = \sum_{e \in E} \omega_e \alpha(e) \beta(e),$$

with positive vertex weights $m_i > 0$ and positive edge weights $\omega_e > 0$. The resulting strongly defined Laplacian acting on 0-forms (functions) takes the form

$$(\mathbb{L}u)_i = \frac{1}{m_i} \sum_{j \sim i} \omega_{ij} (u_i - u_j)$$

and its associated weakly defined cousin is the Laplacians on graphs defined in (3.1). In particular, if $\omega_e = 1$, one recovers the graph Laplacian as the weak version.

3.5. Hodge decomposition. Given a choice of inner products for k-forms on simplicial manifolds, one always obtains a discrete Hodge decomposition. Indeed, for every discrete k-form α one has

$$\alpha = \delta \mu + \delta^* \nu + h,$$

where μ is a $(k-1)$-from, ν is a $(k+1)$-form and h is a harmonic k-from ($\mathbb{L}h = 0$). As in the smooth case, this decomposition is unique and orthogonal with respect to the inner products on k-forms,

$$0 = (\delta \mu, \delta^* \nu)_k = (h, \delta \mu)_k = (h, \delta^* \nu)_k,$$

which immediately follows from $\delta \circ \delta = 0$ and the fact that harmonic forms satisfy $\delta h = \delta^* h = 0$.

Akin to the smooth case, the Hodge decomposition establishes relations to global properties of simplicial manifolds, since the linear space of harmonic k-forms is isomorphic to the k-th simplicial cohomology of the simplicial manifold M. Again, as an application of this fact, consider a compact simplicial surface without boundary. Then the dimension of the space of harmonic 1-forms is equal to twice the genus of the surface—independent of the concrete choice of inner products on k-forms.

3.6. The cotan Laplacian. We conclude the discussion of Laplacians on simplicial manifolds by providing an important example of inner products. In [**Whi57**], Whitney constructs a map from simplicial k-forms (k-cochains) to piecewise linear differential k-forms. In a nutshell, the idea is to linearly interpolate simplicial k-forms across full-dimensional cells. As the simplest example, consider linear interpolation of 0-forms (functions) on vertices. This kind of interpolation can be extended to arbitrary k-forms. The resulting map

$$W : C^k \to L^2\Lambda^k$$

takes simplicial k-forms to square-integrable k-forms on the simplicial manifold, where we assume each simplex to carry the standard Euclidean structure. The Whitney map is the right inverse of the so-called de Rham map,

$$\alpha(\sigma) = \int_\sigma W(\alpha)$$

for all discrete k-forms α and all k-cells σ. For details we refer to [**Whi57**]. The Whitney map W is a chain map—it commutes with the differential ($dW = W\delta$) and thus factors to cohomology.

Dodziuk and Patodi [**DP76**] use the Whitney map in order to define an inner product on discrete k-forms (k-cochains) by

$$(\alpha, \beta)_k := \int_M g(W\alpha, W\beta)\mathrm{vol}_g,$$

where (in our case) g denotes a piecewise Euclidean metric on the simplicial manifold. From the perspective of the *Finite Element Method* (FEM), Whitney's construction is a special case of constructing stable finite elements; see [**AFW06**].

For triangle meshes the resulting strongly defined Laplacian acting on 0-forms (functions) takes the form

$$\mathbb{L} = \mathbb{M}_0^{-1}L,$$

where $\mathbb{M}_0$ is the *mass matrix* given by

$$(\mathbb{M}_0)_{ij} := \begin{cases} \frac{A_{ij}}{12} & \text{if } i \sim j, \\ \frac{A_i}{6} & \text{if } i = j, \\ 0 & \text{otherwise.} \end{cases}$$

Here A_{ij} denotes the combined area of the two triangles incident to edge (i,j) and A_i is the combined area of all triangles incident to vertex i. The corresponding weakly defined Laplacian L is the so-called *cotan Laplace* matrix with entries

$$L_{ij} := \begin{cases} -\frac{1}{2}\left(\cot\alpha_{ij} + \cot\beta_{ij}\right) & \text{if } i \sim j, \\ -\sum_{j\sim i} L_{ij} & \text{if } i = j, \\ 0 & \text{otherwise,} \end{cases}$$

where α_{ij} and β_{ij} are the two angles opposite to edge (i,j).

The cotan Laplacian has been rediscovered many times in different contexts [**Duf59, Dzi88, PP93**]; the earliest explicit mention seems to go back to MacNeal [**Mac49**], but perhaps it was already known at the time of Courant. The cotan Laplacian has been enjoying a wide range of applications in geometry processing (see [**CdGDS13, LZ10, RW14**] and references therein), including barycentric coordinates, mesh parameterization, mesh compression, fairing, denoising, spectral fingerprints, shape clustering, shape matching, physical simulation of thin structures, and geodesic distance computation.

The construction of discrete Laplacians based on inner products can be extended from simplicial surfaces to meshes with (not necessarily planar) polygonal faces [**AW11**, **BLS05a**, **BLS05b**]. The cotan Laplacian has furthermore been extended to semi-discrete surfaces [**Car16**] as well as to subdivision surfaces [**dGDMD16**].

3.7. Desiderata for 'perfect' discrete Laplacians. Structural properties of discrete Laplacians play an important role in applications to geometry processing, e.g., when solving the ubiquitous Poisson problem

$$\mathbb{L}u = f$$

for a given right hand side f and an unknown function u with Dirichlet boundary conditions. For solving this problem, one often prefers to work with the weak formulation ($Lu = \mathbb{M}_0 f$) instead of the strong one ($\mathbb{L}u = f$) since the weakly defined cousins of strongly defined Laplacians satisfy properties SYM and PSD, which allows for efficient linear solvers.

Which properties are desirable for discrete Laplacians on top of SYM and PSD? Moreover, is it possible to maintain *all* properties of smooth Laplacians in the discrete case? To answer this question, we follow [**WMKG07**].

Smooth Laplacians are differential operators that act *locally*. Locality can be represented in the discrete case by working with (weakly defined) discrete Laplacians based on edge weights:

(LOC) vertices i and j do not share an edge $\Rightarrow \omega_{ij} = 0$.

This property reflects locality of action by ensuring that if vertices i and j are not connected by an edge, then changing the function value u_j at a vertex j does not alter the value $(Lu)_i$ at vertex i. Property LOC results in sparse matrices, which can be treated efficiently in computations.

In the smooth setting, linear functions on $\mathbb{R}^2$ are in the kernel of the standard Laplacian on Euclidean domains. For discrete Laplacians on a graph Γ, this property translates into requiring that $(Lu)_i = 0$ at each interior vertex whenever Γ is embedded into the plane with straight edges and u is a *linear* function on the plane, point-sampled at the vertices of Γ, i.e.,

(LIN) $\Gamma \subset \mathbb{R}^2$ embedded and $u : \mathbb{R}^2 \to \mathbb{R}$ linear $\Rightarrow (Lu)_i = 0$ at interior vertices.

In applications, this *linear precision* property is desirable for de-noising of surface meshes [**DMSB99**] (where one expects to remove normal noise only but not to introduce tangential vertex drift), mesh parameterization [**FH05**, **HS17**] (where one expects planar regions to remain invariant under parameterization), and plate bending energies [**RW14**] (which must vanish for flat configurations).

Furthermore, it is often natural and desirable to require positive edge weights:

(POS) vertices i and j share an edge $\Rightarrow \omega_{ij} > 0$.

This requirement implies PSD and is a sufficient (but not a necessary) condition for a *discrete maximum principle* MAX. In diffusion problems corresponding to $u_t = -\Delta u$, POS ensures that flow travels from regions of higher potential to regions of lower potential.

The combination LOC+SYM+POS is related to Tutte's embedding theorem for planar graphs [**GGT06, Tut63**]: Positive weights associated to edges yield a straight-line embedding of an abstract planar graph for a fixed convex boundary polygon. Tutte's embedding is unique for a given set of positive edge weights, and it satisfies LIN by construction since each interior vertex (and therefore its x- and y-coordinate) is a convex combination of its adjacent vertices with respect to the given edge weights.

3.8. No free lunch. Given an arbitrary simplicial mesh, does there exist a discrete Laplacian that satisfies *all* of the desirable properties LOC, SYM, POS, and LIN? Let us start by asking this question for the discrete Laplacians considered so far.

Perhaps the simplest case is to consider the graph Laplacian ($\omega_{ij} = 1$). This Laplacian clearly satisfies LOC+SYM+POS, but in general fails to satisfy LIN due its indifference to the geometry of a graph's embedding.

Next, consider the cotan Laplacian. The edge weights of the cotan Laplacian turn out to be a special case of weights arising from *orthogonal duals.* In particular, edge weights of the cotan Laplacian arise by considering the orthogonal dual obtained by connecting circumcenters of (primal) triangles of a planar triangulation by straight edges; see Figure 1 (left). More generally, consider a graph embedded into the plane with straight edges that do not cross. An orthogonal dual is a realization of the dual graph in the plane, with straight dual edges that are orthogonal to their corresponding primal ones. Different from primal edges, dual edges are allowed to cross each other. Together, a primal graph and its orthogonal dual determine edge weights (on primal edges) defined as the ratio between the *signed* lengths of dual edges and the unsigned lengths of primal edges,

$$\omega_e = \frac{|\star e|}{|e|}.$$

Here, $|e|$ denotes the usual Euclidean length, whereas $|\star e|$ denotes the *signed* Euclidean length of the dual edge. The sign is obtained as follows. The dual edge $\star e$ connects two dual vertices $\star f_1$ and $\star f_2$, corresponding to the primal faces f_1 and f_2, respectively. The sign of $|\star e|$ is positive if along the direction of the ray from $\star f_1$ to $\star f_2$, the primal face f_1 lies before f_2. The sign is negative otherwise. For the special case of duals that arise from connecting circumcenters of a triangulation of the plane, one obtains the cotan weights.

Edge weights obtained from orthogonal duals give rise to discrete Laplacians that satisfy LOC+SYM+LIN. Indeed, while LOC and SYM are immediate by construction, LIN is equivalent to dual edges forming a *closed* polygon (dual face) per primal vertex. To see this equivalence, consider the x- and y- Euclidean coordinates of primal vertices, considered as linear functions over the plane. These linear functions (and therefore all linear functions) are in the kernel of the discrete Laplacian arising from edge weights obtained from orthogonal duals if and only if dual edges form closed polygons around all inner primal vertices. In fact, this equivalence can be reformulated in terms of a century-old result by Maxwell and Cremona [**Cre90, Max64**]: Regard the primal graph as a stress framework with positive edge weights corresponding to contracting edges and negative edge weights regarded as expanding edges. Then the stress framework is in static equilibrium if and only if it satisfies LIN (which constitutes the Euler-Lagrange equations for the

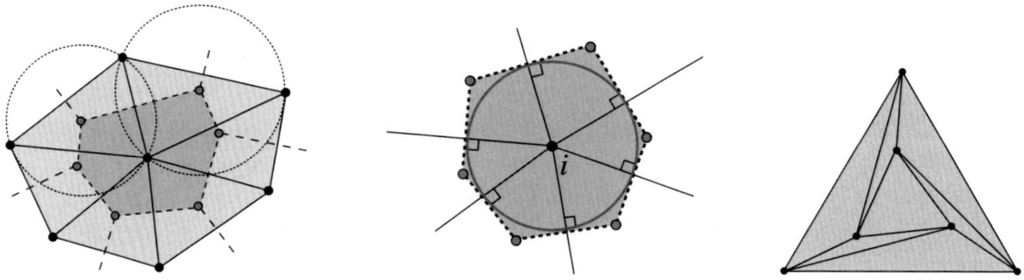

FIGURE 1. *Left*: Primal graph (solid lines) and orthogonal circumcentric dual graph (dashed lines) defining the cotan Laplacian. *Middle*: Mean value weights correspond to dual edges tangent to the unit circle around a primal vertex. *Right*: The projection of the Schönhardt polytope does not allow for a discrete Laplacian satisfying SYM+LOC+LIN+POS.

equilibrium) and thus if and only if there exists an orthogonal dual network that gives rise to the given primal edge weights.

While properties LOC+SYM+LIN are always satisfied by discrete Laplacians arising from orthogonal duals, these Laplacians fail to satisfy POS in general. In fact, so-called *weighted Delaunay triangulations* turn out to be the only triangulations that give rise to positive edge weights arising from orthogonal duals and thus admit discrete Laplacians that satisfy LOC+SYM+LIN+POS. For example, for the cotan Laplacian one has positive edge weights $(\cot\alpha_{ij} + \cot\beta_{ij}) > 0$ if and only if $(\alpha_{ij} + \beta_{ij}) < \pi$. This is the case if and only if the triangulation is Delaunay. As a consequence, if one starts with a triangulation of the plane that is not Delaunay, then the cotan Laplacian fails to satisfy POS. One may restore POS by successive edge flips (thereby changing the combinatorics of the triangulation that one started with) until one arrives at a Delaunay triangulation [**BS07**]. Unfortunately, the number of required edge flips to obtain a Delaunay triangulation from an arbitrary given triangulation cannot be bounded a priori. Therefore, while the approach of edge flips yields discrete Laplacians satisfying SYM+LIN+POS, it fails to yield Laplacian that satisfy LOC in general.

Like Rippa's theorem discussed above, the relation between discrete Laplacians and weighted Delaunay triangulations provides an instance of the intricate connection between properties of discrete differential operators and purely geometric properties.

For completeness, in order to illustrate that there indeed *are* discrete Laplacians that satisfy any choice of three but not all four of the desired properties LOC+SYM+LIN+POS, consider dropping the requirement of symmetry of edge weights. In this case, one enters the realm of barycentric coordinates [**HS17**], where one may still obtain an orthogonal *dual face per primal vertex*, but these dual faces no longer fit together to form a consistent dual graph; see Figure 1 (middle). In particular, for the case dual edges with positive lengths, one obtains edge weights satisfying LOC+LIN+POS but not SYM. Floater et al. [**FHK06**] explored a subspace of this case: a one-parameter family of linear precision barycentric coordinates, including the widely used mean value and Wachspress coordinates. Langer et al. [**LBS06**] showed that each member of this family corresponds to a specific choice of orthogonal dual face per primal vertex.

Summing up, for general simplicial meshes there exists no discrete Laplace operator that satisfies all of the desired properties LOC+SYM+LIN+POS simultaneously; see 1 (right) for a simple example of a mesh that does not admit such a 'perfect' discrete Laplacian. This limitation provides a taxonomy on existing literature and explains the plethora of existing discrete Laplacians: Since not all desired properties can be fulfilled simultaneously, it depends on the application at hand to design discrete Laplacians that are tailored towards the specific needs of a concrete problem.

3.9. Convergence. Another important desideratum is *convergence*: In the limit of refinement of simplicial manifolds that approximate a smooth manifold, one seeks to approximate the smooth Laplacian by a sequence of discrete ones. For applications this is important in terms of obtaining discrete operators that are as mesh-independent as possible—re-meshing a given shape should not result in a drastically different Laplacian.

A closely related concept to convergence is *consistency*. A sequence of discrete Laplacians $(\Delta_n)_{n\in\mathbb{N}}$ is called consistent, if $\Delta_n u \to \Delta u$ for all appropriately chosen functions u. For example, it can be shown that Laplacians on point clouds, such as those considered in [**BSW09**] are consistent; see [**DRW10**].

Convergence is more difficult to show than consistency since it additionally requires that that the solutions u_n to the Poisson problems $\Delta_n u_n = f$ converge (in an appropriate norm) to the solution u of $\Delta u = f$. Discussing convergence in detail is beyond the scope of this short survey. Roughly speaking, Laplacians on simplicial manifolds converge to their smooth counterparts (in an appropriate operator norm) if the inner products on discrete k-forms used for defining simplicial Laplacians converge to the inner products on smooth k-forms. In this case, one obtains convergence of solutions to the Poisson problem, convergence of the components of the Hodge decomposition, convergence of eigenvalues [**DP76, Dzi88, HPW06, War, Wil07**], and (using different techniques) convergence of Cheeger cuts [**TSvB$^+$16**].

References

[AFW06] Douglas N. Arnold, Richard S. Falk, and Ragnar Winther, *Finite element exterior calculus, homological techniques, and applications*, Acta Numer. **15** (2006), 1–155, DOI 10.1017/S0962492906210018. MR2269741

[AK98] Vladimir I. Arnold and Boris A. Khesin, *Topological methods in hydrodynamics*, Applied Mathematical Sciences, vol. 125, Springer-Verlag, New York, 1998. MR1612569

[AW11] Marc Alexa and Max Wardetzky, *Discrete Laplacians on general polygonal meshes*, ACM Transactions on Graphics **30** (2011), no. 4, 102:1–102:10.

[Ber03] Marcel Berger, *A panoramic view of Riemannian geometry*, Springer-Verlag, Berlin, 2003. MR2002701

[BLS05a] Franco Brezzi, Konstantin Lipnikov, and Mikhail Shashkov, *Convergence of the mimetic finite difference method for diffusion problems on polyhedral meshes*, SIAM J. Numer. Anal. **43** (2005), no. 5, 1872–1896, DOI 10.1137/040613950. MR2192322

[BLS05b] Franco Brezzi, Konstantin Lipnikov, and Valeria Simoncini, *A family of mimetic finite difference methods on polygonal and polyhedral meshes*, Math. Models Methods Appl. Sci. **15** (2005), no. 10, 1533–1551, DOI 10.1142/S0218202505000832. MR2168945

[Bro97] Robert Brooks, *Isospectral graphs and isospectral surfaces*, Séminaire de Théorie Spectrale et Géométrie, No. 15, Année 1996–1997, Sémin. Théor. Spectr. Géom., vol. 15, Univ. Grenoble I, Saint-Martin-d'Hères, [1997], pp. 105–113, DOI 10.5802/tsg.184. MR1604259

[BS07] Alexander I. Bobenko and Boris A. Springborn, *A discrete Laplace-Beltrami operator for simplicial surfaces*, Discrete Comput. Geom. **38** (2007), no. 4, 740–756, DOI 10.1007/s00454-007-9006-1. MR2365833

[BSW09] Mikhail Belkin, Jian Sun, and Yusu Wang, *Constructing Laplace operator from point clouds in* $\mathbb{R}^d$, Proceedings of the Twentieth Annual ACM-SIAM Symposium on Discrete Algorithms, SIAM, Philadelphia, PA, 2009, pp. 1031–1040. MR2807545

[Bus82] Peter Buser, *A note on the isoperimetric constant*, Ann. Sci. École Norm. Sup. (4) **15** (1982), no. 2, 213–230. MR683635

[Can13] Yaiza Canzani, *Analysis on manifolds via the Laplacian*, 2013, Lecture notes, Harvard University.

[Car16] Wolfgang Carl, *A Laplace operator on semi-discrete surfaces*, Found. Comput. Math. **16** (2016), no. 5, 1115–1150, DOI 10.1007/s10208-015-9271-y. MR3552842

[CdGDS13] Keenan Crane, Fernando de Goes, Mathieu Desbrun, and Peter Schröder, *Digital geometry processing with discrete exterior calculus*, ACM SIGGRAPH 2013 Courses, SIGGRAPH '13, ACM, 2013.

[Cha84] Isaac Chavel, *Eigenvalues in Riemannian geometry*, Pure and Applied Mathematics, vol. 115, Academic Press, Inc., Orlando, FL, 1984. Including a chapter by Burton Randol; With an appendix by Jozef Dodziuk. MR768584

[Chu07] Fan Chung, *Four proofs for the Cheeger inequality and graph partition algorithms*, Proceedings of the ICCM, 2007.

[Cre90] Luigi Cremona, *Graphical statics (transl. of Le figure reciproche nelle statica graphica, milano, 1872)*, Oxford University Press, 1890.

[CXGL10] Renjie Chen, Yin Xu, Craig Gotsman, and Ligang Liu, *A spectral characterization of the Delaunay triangulation*, Comput. Aided Geom. Design **27** (2010), no. 4, 295–300, DOI 10.1016/j.cagd.2010.02.002. MR2607756

[dGDMD16] Fernando de Goes, Mathieu Desbrun, Mark Meyer, and Tony DeRose, *Subdivision exterior calculus for geometry processing*, ACM Transactions on Graphics **35** (2016), no. 4, 133:1–133:11.

[DHLM05] Mathieu Desbrun, Anil Hirani, Melvin Leok, and Jerrold E. Marsden, *Discrete Exterior Calculus*, arXiv:math.DG/0508341.

[DMSB99] Mathieu Desbrun, Mark Meyer, Peter Schröder, and Alan H. Barr, *Implicit fairing of irregular meshes using diffusion and curvature flow*, Proceedings of ACM SIGGRAPH, 1999, pp. 317–324.

[DP76] J. Dodziuk and V. K. Patodi, *Riemannian structures and triangulations of manifolds*, J. Indian Math. Soc. (N.S.) **40** (1976), no. 1-4, 1–52 (1977). MR488179

[DRW10] Tamal K. Dey, Pawas Ranjan, and Yusu Wang, *Convergence, stability, and discrete approximation of Laplace spectra*, Proceedings of the Twenty-First Annual ACM-SIAM Symposium on Discrete Algorithms, SIAM, Philadelphia, PA, 2010, pp. 650–663. MR2768624

[Duf59] R. J. Duffin, *Distributed and lumped networks*, J. Math. Mech. **8** (1959), 793–826, DOI 10.1512/iumj.1959.8.58051. MR0106032

[Dzi88] Gerhard Dziuk, *Finite elements for the Beltrami operator on arbitrary surfaces*, Partial differential equations and calculus of variations, Lecture Notes in Math., vol. 1357, Springer, Berlin, 1988, pp. 142–155, DOI 10.1007/BFb0082865. MR976234

[Eva98] Lawrence C. Evans, *Partial differential equations*, Graduate Studies in Mathematics, vol. 19, American Mathematical Society, Providence, RI, 1998. MR1625845

[FH05] Michael S. Floater and Kai Hormann, *Surface parameterization: a tutorial and survey*, Advances in multiresolution for geometric modelling, Math. Vis., Springer, Berlin, 2005, pp. 157–186, DOI 10.1007/3-540-26808-1_9. MR2112350

[FHK06] Michael S. Floater, Kai Hormann, and Geza Kós, *A general construction of barycentric coordinates over convex polygons*, Adv. Comp. Math. **24** (2006), no. 1–4, 311–331.

[Flo03] Michael S. Floater, *Mean value coordinates*, Comput. Aided Geom. Design **20** (2003), no. 1, 19–27, DOI 10.1016/S0167-8396(03)00002-5. MR1968304

[Fra04] Theodore Frankel, *The geometry of physics*, 2nd ed., Cambridge University Press, Cambridge, 2004. An introduction. MR2060795

[GGT06] Steven J. Gortler, Craig Gotsman, and Dylan Thurston, *Discrete one-forms on meshes and applications to 3D mesh parameterization*, Comput. Aided Geom. Design **23** (2006), no. 2, 83–112, DOI 10.1016/j.cagd.2005.05.002. MR2189438

[GWW92] Carolyn Gordon, David L. Webb, and Scott Wolpert, *One cannot hear the shape of a drum*, Bull. Amer. Math. Soc. (N.S.) **27** (1992), no. 1, 134–138, DOI 10.1090/S0273-0979-1992-00289-6. MR1136137

[HPW06] Klaus Hildebrandt, Konrad Polthier, and Max Wardetzky, *On the convergence of metric and geometric properties of polyhedral surfaces*, Geom. Dedicata **123** (2006), 89–112, DOI 10.1007/s10711-006-9109-5. MR2299728

[HS17] Kai Hormann and N. Sukumar (eds.), *Generalized barycentric coordinates in computer graphics and computational mechanics*, CRC Press, Boca Raton, FL, 2018. MR3752645

[JMD+07] Pushkar Joshi, Mark Meyer, Tony DeRose, Brian Green, and Tom Sanocki, *Harmonic coordinates for character articulation*, ACM Trans. Graph. **26** (2007), no. 3, 71:1–71:10.

[JSW05] Tao Ju, Scott Schaefer, and Joe Warren, *Mean value coordinates for closed triangular meshes*, ACM Trans. Graph. **24** (2005), no. 3, 561–566.

[Kac66] Mark Kac, *Can one hear the shape of a drum?*, Amer. Math. Monthly **73** (1966), no. 4, 1–23, DOI 10.2307/2313748. MR201237

[KVV04] Ravi Kannan, Santosh Vempala, and Adrian Vetta, *On clusterings: good, bad and spectral*, J. ACM **51** (2004), no. 3, 497–515, DOI 10.1145/990308.990313. MR2145863

[LBS06] Torsten Langer, Alexander Belyaev, and Hans-Peter Seidel, *Spherical barycentric coordinates*, Siggraph/Eurographics Sympos. Geom. Processing, 2006, pp. 81–88.

[LZ10] Bruno Lévy and Hao (Richard) Zhang, *Spectral mesh processing*, ACM SIGGRAPH 2010 Courses, SIGGRAPH '10, ACM, 2010.

[Mac49] R. MacNeal, *The Solution of Partial Differential Equations by Means of Electrical Networks*, 1949, PhD thesis, California Institute of Technology.

[Max64] J. C. Maxwell, *On reciprocal figures and diagrams of forces*, Phil. Mag. **27** (1864), 250–261.

[PP93] Ulrich Pinkall and Konrad Polthier, *Computing discrete minimal surfaces and their conjugates*, Experiment. Math. **2** (1993), no. 1, 15–36. MR1246481

[Rip90] Samuel Rippa, *Minimal roughness property of the Delaunay triangulation*, Comput. Aided Geom. Design **7** (1990), no. 6, 489–497, DOI 10.1016/0167-8396(90)90011-F. MR1079398

[Ros97] Steven Rosenberg, *The Laplacian on a Riemannian manifold*, London Mathematical Society Student Texts, vol. 31, Cambridge University Press, Cambridge, 1997. An introduction to analysis on manifolds. MR1462892

[RW14] Martin Rumpf and Max Wardetzky, *Geometry processing from an elastic perspective*, GAMM-Mitt. **37** (2014), no. 2, 184–216, DOI 10.1002/gamm.201410009. MR3285020

[Sch95] Günter Schwarz, *Hodge decomposition—a method for solving boundary value problems*, Lecture Notes in Mathematics, vol. 1607, Springer-Verlag, Berlin, 1995. MR1367287

[Sun85] Toshikazu Sunada, *Riemannian coverings and isospectral manifolds*, Ann. of Math. (2) **121** (1985), no. 1, 169–186, DOI 10.2307/1971195. MR782558

[Sun07] Toshikazu Sunada, *Discrete geometric analysis*, Analysis on graphs and its applications, Proc. Sympos. Pure Math., vol. 77, Amer. Math. Soc., Providence, RI, 2008, pp. 51–83, DOI 10.1090/pspum/077/2459864. MR2459864

[TSvB+16] Nicolás García Trillos, Dejan Slepčev, James von Brecht, Thomas Laurent, and Xavier Bresson, *Consistency of Cheeger and ratio graph cuts*, J. Mach. Learn. Res. **17** (2016), Paper No. 181, 46. MR3567449

[Tut63] W. T. Tutte, *How to draw a graph*, Proc. London Math. Soc. (3) **13** (1963), 743–767, DOI 10.1112/plms/s3-13.1.743. MR0158387

[War] Max Wardetzky, *Convergence of the cotangent formula: an overview*, Discrete differential geometry, Oberwolfach Semin., vol. 38, Birkhäuser, Basel, 2008, pp. 275–286, DOI 10.1007/978-3-7643-8621-4_15. MR2405672

[Whi57] Hassler Whitney, *Geometric integration theory*, Princeton University Press, Princeton, N. J., 1957. MR0087148

[Wil07] Scott O. Wilson, *Cochain algebra on manifolds and convergence under refinement*, Topology Appl. **154** (2007), no. 9, 1898–1920, DOI 10.1016/j.topol.2007.01.017. MR2319262

[WMKG07] Max Wardetzky, Saurabh Mathur, Felix Kälberer, and Eitan Grinspun, *Discrete Laplace operators: No free lunch*, Siggraph/Eurographics Sympos. Geom. Processing, 2007, pp. 33–37.

[ZGLG12] Wei Zeng, Ren Guo, Feng Luo, and Xianfeng Gu, *Discrete heat kernel determines discrete Riemannian metric*, Graphics Models **74** (2012), no. 4, 121–129.

Institute of Applied and Numerical Mathematics, Georg-August-University Göttingen, Germany

Email address: `wardetsky@math.uni-goettingen.de`

Proceedings of Symposia in Applied Mathematics
Volume **76**, 2020
https://doi.org/10.1090/psapm/076/00670

Discrete parametric surfaces

Johannes Wallner

1. Introduction

Discrete parametric surfaces are discrete analogues of smooth parametric surfaces. They are, however, not simply discrete approximations of their smooth counterparts, but are the subject of a separate discrete theory. As it turns out, a systematic theory of parametric surfaces can be based on integrable systems, and the discrete case can be interpreted as a "master" case which contains smooth surfaces as a limit.

This section on discrete parametric surfaces is organized as follows: We first introduce notation. Two particular kinds of discrete surfaces are discussed next: circular nets in §2, and K-nets in §3 are examples of a 3-system and a 2-system, respectively. In the case of K-nets, we also discuss the relation to the sine-Gordon equation. We then show applications within mathematics in §4, cf. [**BHS06**], and the connection with freeform architecture in §5, cf. [**PW16**]. The main source for this chapter is the monograph [**BS09**].

1.1. Discrete, semidiscrete, and continuous surfaces. The simplest case of a continuous surface is a point $x(u_1, u_2)$ of space depending on two parameters u_1, u_2. A discrete surface $x(i_1, i_2)$ is the same, only the parameters i_1, i_2 are integers. The notion of *transformation* of a surface usually refers to a pair $x(u_1, u_2)$ and $x^{(1)}(u_1, u_2)$ of surfaces which are in a certain relation (images taken from [**BS09**]):

(1)

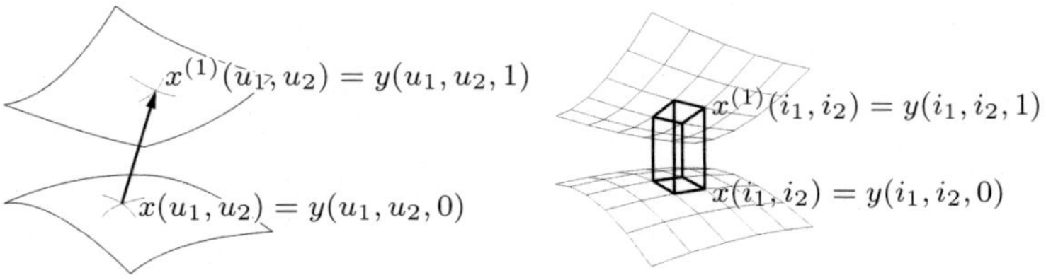

The theory of transformations was fully developed in the 1920s, cf. [**Eis23**]. Well known instances are Darboux transforms, Bäcklund transforms, or the Christoffel transform (which relates minimal surfaces with conformal mappings into sphere,

2020 *Mathematics Subject Classification.* Primary 37K25; Secondary 53A05.

Key words and phrases. Discrete differential geometry, integrable systems.

The author was supported by by SFB-Transregio Program *Discretization in Geometry and Dynamics*, Austrian Science Fund grant no. I2978.

and leads to the Weierstrass representation of minimal surfaces in terms of meromorphic functions). A sequence of surfaces created by iterated application of a transformation rule constitutes a *semidiscrete* object $y(u_1, u_2, i_3)$, where i_3 is an integer parameter:

$$y(u_1, u_2, 0) = x(u_1, u_2), \quad y(u_1, u_2, 1) = x^{(1)}(u_1, u_2), \quad y(u_1, u_2, 2) = x^{(11)}(u_1, u_2), \ \dots$$

Transformations of discrete surfaces can be treated in a similar way: With

$$y(i_1, i_2, 0) = x(i_1, i_2), \quad y(i_1, i_2, 1) = x^{(1)}(i_1, i_2), \quad y(i_1, i_2, 2) = x^{(11)}(i_1, i_2), \ \dots$$

we define a mapping y from $\mathbb{Z}^3$ to space. We see that a sequence of discrete k-dimensional surfaces is nothing but a discrete $(k+1)$-dimensional surface, and the special role of the first two parameters disappears.

A particular feature of many transformations is that they enjoy *Bianchi permutability*: If x has transforms $x^{(1)}$ and $x^{(2)}$, then there exists $x^{(12)}$ which is a transform of both $x^{(1)}$ and $x^{(2)}$ simultaneously (image taken from [**BS09**]):

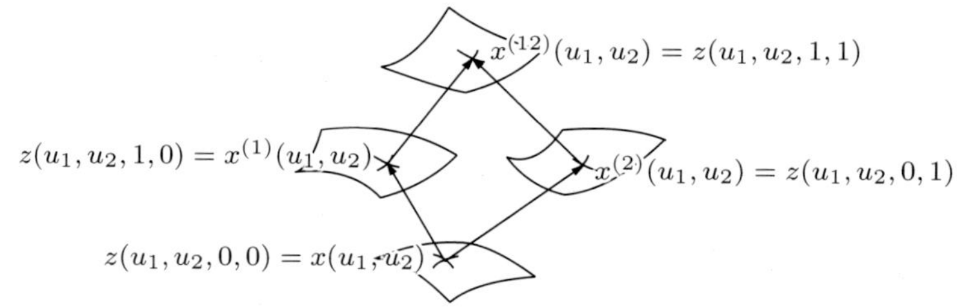

(2)

By iterating this procedure we create a two-dimensional lattice of surfaces, i.e., a mapping "z" from $\mathbb{R}^2 \times \mathbb{Z}^2$ to space. It can be seen as a semidiscrete 4-dimensional surface $z(u_1, u_2, i_3, i_4)$. It is very interesting that the 2D discrete surfaces $x(i_1, i_2) = z(u_1, u_2, i_3, i_4)$ contained in this 4-dimensional semidiscrete object typically exhibit geometric properties similar to discrete surfaces: *The transformations which apply to a class of smooth surfaces provide guidelines on how to find a class of analogous discrete surfaces.*

However if x is a discrete surface to begin with, a 2-dimensional lattice of surfaces is nothing but a 4-dimensional discrete surface $z(i_1, i_2, i_3, i_4)$. The special role of the first two parameters disappears.

It is a key principle of discrete differential geometry that the smooth theory can be obtained by a limit process from the discrete one. We can let a 2D discrete surface converge to a smooth surface, a 3D discrete surface to a sequence of smooth surfaces, and so on.

1.2. History. Integrable equations. An important feature of parametric surfaces is the relation to integrable equations. This connection has a long tradition. The best known example concerns a special "asymptotic" parametrization of surfaces whose Gauss curvature equals the constant -1. It turns out that the angle $\phi(u_1, u_2)$ between parameter lines obeys the sine-Gordon equation

$$\partial_{12}\phi = \sin\phi. \tag{3}$$

It is a fact that for any solution $\phi(u_1, u_2)$ of this equation, another solution $\phi^{(1)}$, called the *Bäcklund transform* of ϕ, can be defined by

$$\partial_1\phi^{(1)} = \partial_1\phi + 2a\sin\frac{\phi + \phi^{(1)}}{2}, \ \partial_2\phi^{(1)} = -\partial_2\phi + \frac{2}{a}\sin\frac{\phi^{(1)} - \phi}{2}.$$

This Bäcklund transformation of functions corresponds directly to the Bäcklund transformation of surfaces: ϕ and $\phi^{(1)}$ are angle functions associated with a Bäcklund pair of surfaces x and $x^{(1)}$. There is even a Bianchi permutability theorem analogous to the surface case: With the arrow symbolizing the Bäcklund relation, we have

$$\begin{array}{ccc} & \phi^{(1)} & \\ \phi \nearrow & & \\ \phi \searrow & & \\ & \phi^{(2)} & \end{array} \Longrightarrow \exists\, \phi^{(12)} \text{ such that } \begin{array}{ccccc} & \phi^{(1)} & \\ \phi \nearrow & & \searrow \phi^{(12)}. \\ \phi \searrow & & \nearrow \\ & \phi^{(2)} & \end{array} \tag{4}$$

Surfaces of constant Gaussian curvature have been instrumental in the development of the systematic "integrability" theory of discrete surfaces. The original Bäcklund transform of smooth K-surfaces was published in 1883 [**Bäck83**], and the permutablity theorem soon after [**Bia02**]. Discrete K-surfaces were constructed in the 1950's [**Sau50, Wun51**]. The connection of discrete surfaces to the discrete Hirota equation, which is a discrete analogue of the sine-Gordon equation, was revealed by [**BP96**]. We are coming back to this topic in §3.

Typically, for any particular class of surfaces (e.g. the circular nets of §2 or the K-nets of §3) one needs some elementary geometric construction in order to establish multidimensional discrete surfaces. Since the discrete case serves as a master theory from which the continuous one is obtained by a limit process, that elementary geometric theorem can be seen as the discrete analogue of the various kinds of integrability needed to construct surfaces and their transformations.

1.3. Curvatures. Surface classes can be equivalently characterized by different properties. E.g., a surface is minimal $\iff$ it is locally area-minimizing $\iff$ its mean curvature vanishes $\iff$ it has a parametrization in terms of of meromorphic functions $f(z)$ and $g(z)$ such that fg^2 is holomorphic, with $x(u,v) = \frac{1}{2}\operatorname{Re}\int_0^{u+iv}\big(f(z)(1-g(z)^2),\, if(z)(1+g(z)^2),\, 2f(z)g(z)\big)\,dz$.

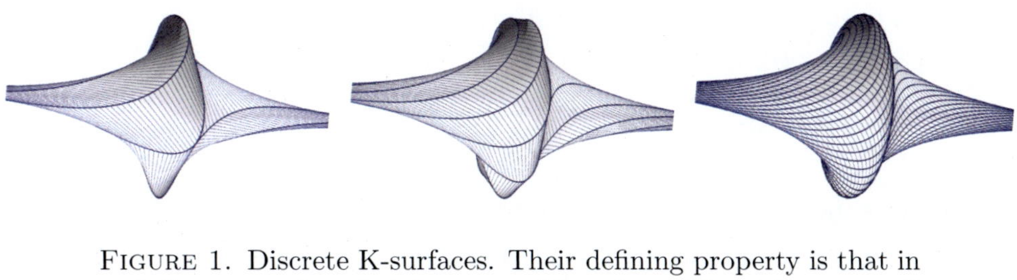

FIGURE 1. Discrete K-surfaces. Their defining property is that in each face, opposite edges have equal length, and in each vertex, all four incident edges are co-planar. Discrete surfaces can converge to semidiscrete or continuous surfaces.

It is this Weierstrass representation which leads to a discretization in the context of *parametric* surfaces (see §4). Area minimization is the basis of a different discretization, which is not parametric [**PP93**].

It is a curious fact that many classes of surfaces which are originally defined via *curvature* (K-surfaces, minimal surfaces, cmc surfaces) had well-established discrete counterparts which were defined by an equivalent characterization not involving curvatures. However, meanwhile curvatures for discrete surfaces have been studied in a systematic way, and there are general concepts of curvatures which apply to the discrete surface classes mentioned above [**BPW10, HSFW17**].

1.4. Discretization principles. The difficulty of assigning curvatures illustrates an important issue: It is not clear a priori which of the various equivalent properties of a class of smooth surfaces should be the one which guides the discretization. However, a discretization is good, or worthy of further investigation, if not just one property carries over from the smooth to the discrete setting, but more than one. For example, the variational definition of discrete minimal surfaces by [**PP93**] leads to simplicial minimal surfaces which not only minimize area, but which also exhibit an associated family of minimal surfaces. The guiding principle of discretization in the case of *parametric* surfaces often is integrability, or (in the discrete case), so-called multidimensional consistency.

1.5. Notation for continuous and discrete nets. A smooth parametric surface maps a parameter value $u \in \mathbb{R}^d$ to a point $x(u) \in \mathbb{R}^n$. Derivatives are described by the symbols

$$\partial_k x, \ \partial_{kl} x, \dots$$

A discrete parametric surface maps an *integer* parameter value $u = (u_1, \dots, u_d) \in \mathbb{Z}^d$ to a point $x(u) \in \mathbb{R}^n$. The role of derivatives is played by differences, e.g.

$$\Delta_1 x(u_1, u_2, \dots, u_d) = x(u_1 + 1, u_2, \dots, u_d) - x(u_1, u_2, \dots, u_d),$$
$$\Delta_2 x(u_1, u_2, \dots, u_d) = x(u_1, u_2 + 1, \dots, u_d) - x(u_1, u_2, \dots, u_d).$$

In general, we use a lower index i to indicate a forward shift in the i-th parameter, and the index $\bar{i}$ to indicate a backward shift.

$$x_k(\dots, x_k, \dots) = x(\dots, u_k + 1, \dots) \qquad x_{\bar{k}}(\dots, x_k, \dots) = x(\dots, u_k - 1, \dots)$$

We use the notation x_{kk}, x_{kl}, $x_{\bar{k}l}$, and so on for iterated shifts. E.g. in a 2-dimensional net $x(u_1, u_2)$ we use a diagram like the one shown in the inset figure to indicate the immediate neighborhood of a general point $x(u_1, u_2)$. The forward difference operator is expressed as $\Delta_k x = x_k - x$.

$$\begin{array}{ccccc} x_{\bar{1}2} & - & x_2 & - & x_{12} \\ | & & | & & | \\ x_{\bar{1}} & - & x & - & x_1 \\ | & & | & & | \\ x_{\bar{1}\bar{2}} & - & x_{\bar{2}} & - & x_{1\bar{2}} \end{array}$$

1.6. Examples. Conjugate surfaces. A *conjugate* parametrization is one where infinitesimal parameter quads are planar. For 2-surfaces, this means that we require

$$\begin{aligned} &3\,\mathrm{vol}\big(\text{c.h.}\,\big(x(u),\ x(u_1+\varepsilon, u_2),\ x(u_1, u_2+\varepsilon),\ x(u_1+\varepsilon, u_2+\varepsilon)\big)\big)/\varepsilon^4 \\ &= \det(x(u_1+\varepsilon, u_2) - x(u),\ x(u_1, u_2+\varepsilon) - x(u),\ x(u_1+\varepsilon, u_2+\varepsilon) - x(u))/2\varepsilon^4 \\ &\approx \det(\varepsilon\partial_1 x,\ \varepsilon\partial_2 x,\ \varepsilon\partial_1 x + \varepsilon\partial_2 x + 2\varepsilon^2\partial_{12} x)/2\varepsilon^4 \\ &= \det(\partial_1 x, \partial_2, \partial_{12} x) = 0 \end{aligned}$$

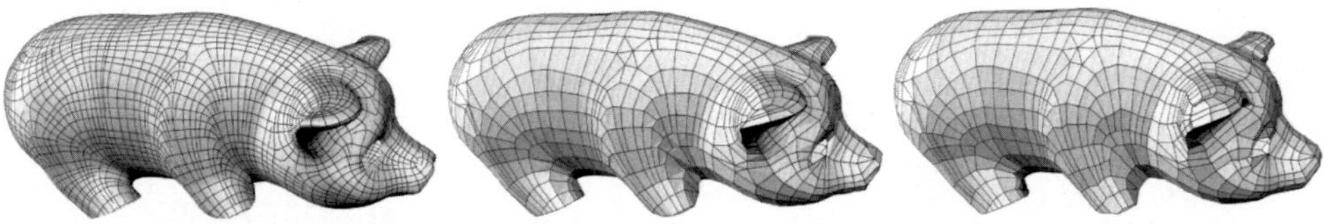

FIGURE 2. *Left:* A conjugate parametric surface $x : U \subset \mathbb{R}^2 \to \mathbb{R}^3$ covers most of this geometric shape. *Center:* discrete parametric surfaces extracted by sampling have elementary quadrilaterals which are almost planar. *Right:* Global optimization makes all faces planar. This optimization problem is numerically feasible because we are already very close to a solution.

The "$\approx$" sign means equality up to remainder terms in the Taylor expansion as $\varepsilon \to 0$. We say that all infinitesimal elementary quadrilaterals of the net are planar. It is obvious how to translate the defining property

$$\det(\partial_k x, \partial_l x, \partial_{kl} x) = 0$$

to the discrete case: a discrete conjugate net is defined by

$$\det(\Delta_k x, \Delta_l x, \Delta_{kl} x) = 0.$$

Generically this is equivalent to

$$\begin{array}{ccc} x_l & — & x_{kl} \\ | & & | \\ x & — & x_k \end{array} \qquad \text{co-planar,} \tag{5}$$

or to existence of coefficients c^{lk}, c^{kl} associated with the elementary quadrilateral $x x_k x_l x_{kl}$, such that

$$\Delta_k \Delta_l x = c^{lk} \Delta_k x + c^{kl} \Delta_l x.$$

For reasons which become apparent later, we use this property as definition of a discrete conjugate surface (a synonym is *conjugate net*).

2. Circular nets: A 3-system

There are two major classes of discrete parametric surfaces guided by the discretization principle of multidimensional consistency. These are the 2-systems and the 3-systems, and we explain these notions by means of prominent examples: the K-nets for the 2-systems, and the circular nets for the 3-systems. We begin with the 3-system case.

DEFINITION 2.1. A circular net is a discrete parametric surface where all elementary quadrilaterals are co-circular, i.e., for every vertex x and neighbors x_k, x_l, there is a circle which contains the four vertices x, x_k, x_l, x_{kl}.

Obviously, a circular net is also conjugate. The following result concerning existence of conjugate nets and circular nets implicitly contains the definition of a 3-system:

THEOREM 2.2. *Conjugacy of nets is a 3-system: Generically, a conjugate net $x(u_1, u_2, u_3)$ is uniquely determined by arbitrary initial values $x(0, u_2, u_3)$ and $x(u_1, 0, u_3)$ and $x(u_1, u_2, 0)$. The same applies to circularity.*

FIGURE 3. *Left:* 2D circular net. *Right:* An elementary cell of a 3D circular net.

PROOF. Consider a combinatorial cube with diagonal x—x_{123}, see the figure below. Assume that the quadrilaterals $xx_1x_2x_{12}$, $xx_1x_3x_{13}$, $xx_2x_3x_{23}$ adjacent to x are planar. Does there exist x_{123} such that the three quads adjacent to x_{123} are planar?

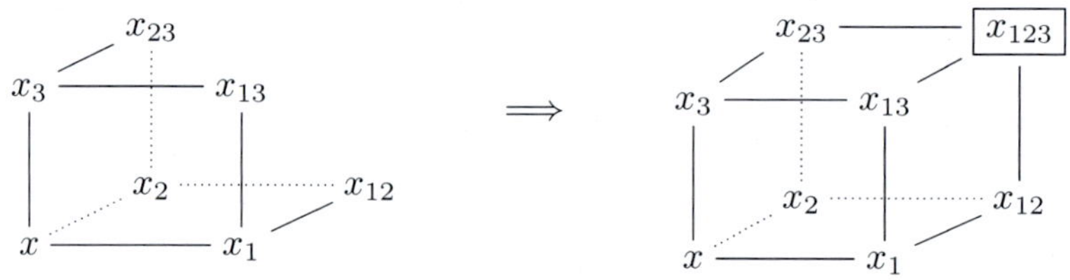

The answer is trivially yes for a conjugate net in $\mathbb{R}P^n$, $n \geq 3$, since the planes carrying the three missing faces are given by three vertices each, and x_{123} is the intersection of these planes.

For circular nets, this is not so obvious: The circumcircles of the three quads incident to x_{123} are given by three points each, but do they intersect in a common point x_{123}? The answer is yes: whenever the three quads $xx_kx_lx_{kl}$ have a circumcircle, then the three circumcircles of x_1, x_{12}, x_{13}, x_2, x_{23}, x_{21}, x_3, x_{31}, x_{32} meet in a common point x_{123}.

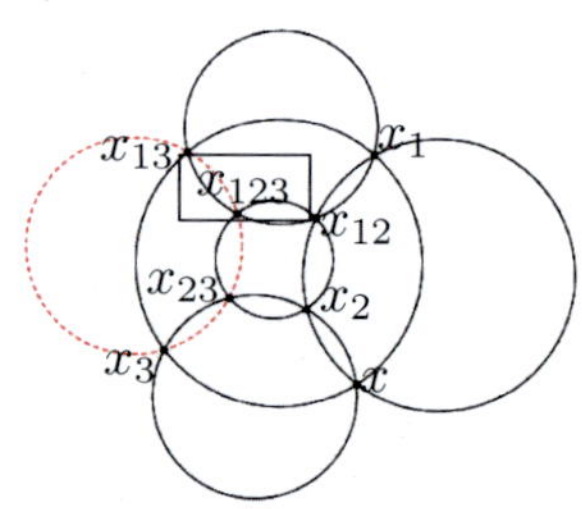

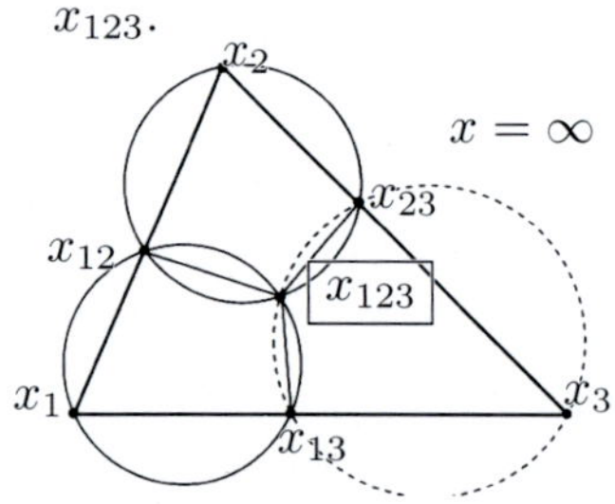

For a proof we observe that the statement is not affected by Möbius transformations. We therefore apply an inversion to move x to infinity. Circles passing through x become straight lines, so x_{kl} becomes a point of the edge x_kx_l of the triangle $x_1x_2x_3$. The statement about circumcircles is now shown by using elementary geometry. □

The proof makes it clear that the "right" geometric setting for conjugate nets is projective space, whereas circular nets should be treated in Möbius geometry.

PROPOSITION 2.3. *Assume that $x(i_1, i_2, i_3)$ is a conjugate net, and that all elementary quadrilaterals which contain vertices with $i_1 = 0$ or $i_2 = 0$ or $i_3 = 0$ are* *circular. Then, generically, circularity propagates through the net: all elementary quads are circular. The same is true for n-dimensional nets with* $n \geq 3$.

PROOF. It is sufficient to show this statement for a 3-cell with diagonal x—x_{123} where the three quads incident to x are circular. If we require circularity, then x_{123} is uniquely determined as the intersection of circumcircles of the quads incident to it. But x_{123} is *already* uniquely determined by the intersection of the planes which carry those quads. It follows that x_{123} is the same regardless if it is computed via the 3-system "conjugacy" or via the 3-system "circularity". □

LEMMA 2.4. *Consider the coefficient functions used in the definition of a conjugate net, see* (5), $\Delta_k \Delta_l x = c^{lk}\Delta_k x + c^{kl}\Delta_l x$. *For a conjugate 3-net, there is a birational mapping*

$$(c^{12}, c^{21}, c^{23}, c^{32}, c^{31}, c^{13}) \overset{\phi}{\longmapsto} (c^{12}_3, c^{21}_3, c^{23}_1, c^{32}_1, c^{31}_2, c^{13}_2)$$

of the coefficients in quads incident to x, *to coefficients in quads incident to* x_{123}.

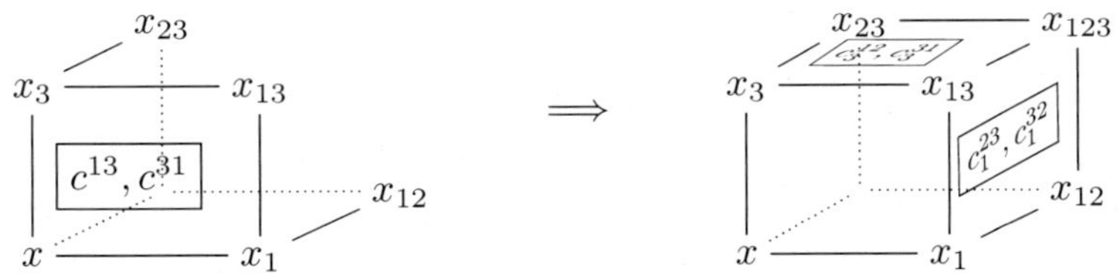

PROOF. We use the "product rule" $\Delta_j(a \cdot b) = a_j \cdot \Delta_j b + \Delta_j a \cdot b$ to expand

$$\begin{aligned}\Delta_i\Delta_j\Delta_k x &= \Delta_i(c^{kj}\Delta_j x + c^{jk}\Delta_k x) = c^{kj}_i \Delta_i\Delta_j x + \Delta_i c^{kj}\Delta_j x + \cdots \\ &= (c^{kj}_i c^{ji} + c^{jk}_i c^{ki})\Delta_i x + (c^{kj}_i c^{ij} + \Delta_i c^{kj})\Delta_j x + (c^{jk}_i c^{ik} + \Delta_i c^{jk})\Delta_k x.\end{aligned}$$

Permuting indices yields

$$\begin{aligned}\Delta_j\Delta_k\Delta_i x &= (c^{ik}_j c^{kj} + c^{ki}_j c^{ij})\Delta_j x + (c^{ik}_j c^{jk} + \Delta_j c^{ik})\Delta_k x + (c^{ki}_j c^{ji} + \Delta_j c^{ki})\Delta_i x. \\ \Delta_k\Delta_i\Delta_j x &= (c^{ji}_k c^{ik} + c^{ij}_k c^{jk})\Delta_k x + (c^{ji}_k c^{ki} + \Delta_k c^{ji})\Delta_i x + (c^{ij}_k c^{kj} + \Delta_k c^{ij})\Delta_j x.\end{aligned}$$

Since $\Delta_i\Delta_j\Delta_k$ is invariant w.r.t. permutations of indices, and generically the first derivatives are linearly independent, we get

$$\Delta_i c^{jk} = c^{ij}_k c^{jk} + c^{ji}_k c^{ik} - c^{jk}_i c^{ik} \quad (i \neq j \neq k \neq i).$$

These are 6 linear equations for the 6 variables $(c^{12}_3, \dots, c^{32}_1)$. Their explicit solution by matrix inversion yields the desired rational mapping ϕ, and analogously for the inverse ϕ^{-1}. Even if ϕ was derived under the assumption of linearly independent first derivatives, it may be applied to all input data. □

The following result implicitly defines an *integrable* discrete 3-system:

THEOREM 2.5. *Conjugacy is a 4-consistent (i.e., integrable) 3-system: If in the combinatorial 4-cube with diagonal* $x \dots x_{1234}$, *conjugacy is imposed on the faces incident to* x *(and by the 3-system property , on the four 3-cubes incident to* x*), then there is a generically unique choice of* x_{1234} *such that also the remaining quads of the hypercube enjoy the property. Analogously, circularity is 4-consistent.*

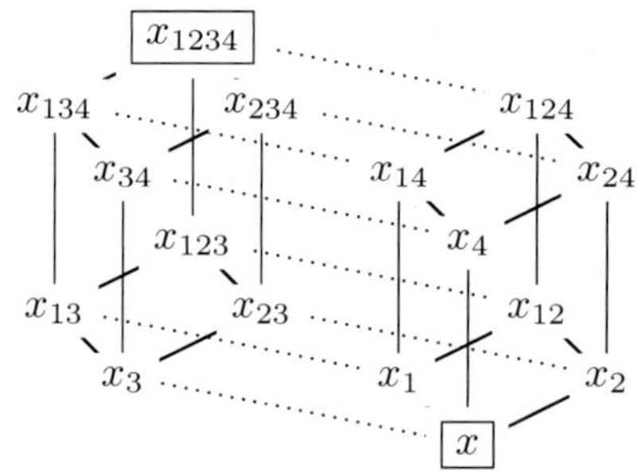

PROOF. The hypercube with diagonal x—x_{1234} contains four 3-cubes incident with x_{1234}. Within each of these cubes, x_{1234} is found as the intersection of 3 quads. E.g. from the cube with diagonal x—x_{123} we get

$$x_{1234} \in \operatorname{span}(x_{12}x_{123}x_{124}) \cap \operatorname{span}(x_{13}x_{132}x_{134}) \cap \operatorname{span}(x_{14}x_{142}x_{143}). \tag{6}$$

The cubes with diagonals x—x_{124}, x—x_{134}, and x—x_{224} yield three more ways to express x_{1234} as intersection of three planes. We must show that all these are equal.

(1) Each of the quads mentioned above is the intersection of two adjacent 3-cubes. Counting shows that these six cubes are actually all four 3-cubes incident with x_{1234}.

(2) Thus, in case of dimension ≥ 4 and general position, (6) is transformed into an intersection of four 3-spaces (each is the span of a 3-cube). Since all 3-cubes incident to x_{1234} occur in this expression, the transformed expression is invariant w.r.t. permutation of indices $\Longrightarrow$ all ways of computing x_{1234} yield the same result.

(3) x_{1234} can also be found in an alternative way, namely by using the coefficients c^{ij} defined by (5). We can compute them in the quads incident to x, apply ϕ, and compute vertices $x_{123}, x_{124}, x_{134}, x_{234}$. Repetition of this procedure for the quads incident with x_{1234} yields four different expressions for x_{1234}, within each of the four cubes incident to x_{1234}, i.e., there are four rational functions of the arguments x, x_i, x_{ij} which in the general position case and dimension ≥ 4 evaluate to the same point x_{1234}, by (2).

(4) Since almost all arguments have the general position property, the rational functions constructed above are equal. They are independent of the dimension, so they can be applied also in 3-space, proving 4-consistency.

Consistency of *circularity* follows as a corollary, since circularity propagates by Prop. 2.3. □

An easy combinatorial argument even shows that 4-consistency of 3-systems implies n-consistency for all $n \geq 4$.

2.1. Principal curvature lines. We have not yet mentioned the smooth class of surfaces which corresponds to the circular nets: It is the parametrization of surfaces along principal curvature lines. There are several reasons for that: Circularity of quads is a discrete version of *orthogonality*, but a more compelling reason is the behavior of the surface normals. A curve $c(t)$ on a surface is a principal curvature line, if progress in that direction causes the unit normal vector to change in the same direction ($\frac{d}{dt}n(c(t)) = \lambda(t)\frac{d}{dt}c(t)$, the tangent vectors of these curves are eigenvectors of the Weingarten map $-dn$), so the surface traced out by the surface normals is developable.

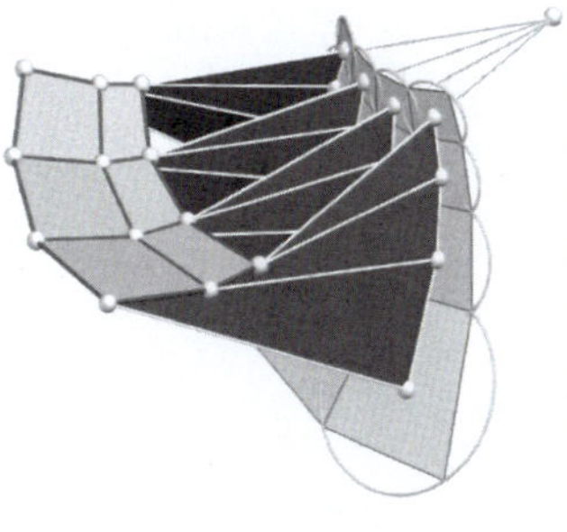

Developability means that a surface normal $N(c(t))$ and its infinitesimal neighbor $N(c(t+dt))$ "intersect" each other (in the same way an infinitesimal quad of a conjugate net is planar). It is this surface of normals which has a proper discrete analogue: As we progress along a sequence of adjacent faces, the successive axes of circles intersect, which yields a discrete developable surface. Further, convergence of circular nets to principal parametrizations can be shown [**BS09**].

3. K-nets: A 2-system

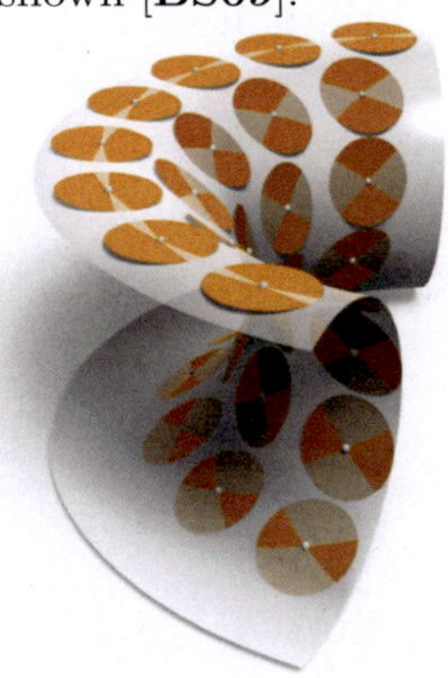

3.1. Asymptotic nets. Asymptotic parametrizations have elementary quads which are as non-planar as possible. For a negatively curved 2-surface in $\mathbb{R}^3$, the asymptotic tangents in a point are found by intersecting the surface with its own tangent plane; the parameter lines of the asymptotic tangents are the integral curves of these tangents. The asymptotic condition reads

$$\partial_{11}x, \partial_{22}x \in \operatorname{span}\{\partial_1 x, \partial_2\}. \tag{7}$$

A discrete version of this condition is the following: All symmetric 2nd differences around a vertex are contained in the plane spanned by the first differences. This can be formulated in a symmetric way:

$$x, x_1, x_{\bar{1}}, x_2, x_{\bar{2}} \in \text{plane } P(u) \tag{8}$$

$$\begin{array}{ccccc} & & x_2 & & \\ & & | & & \\ x_{\bar{1}} & — & x & — & x_1 \\ & & | & & \\ & & x_{\bar{2}} & & \end{array}$$

Consequently, discrete A-nets are defined by the requirement that all edges emanating from a vertex $x(u)$ must lie in a common plane $P(u)$.

3.2. Surfaces of constant Gaussian curvature. Surfaces of constant negative Gaussian curvature (*K-surfaces*) are interesting for a variety of reasons, e.g. because the geometry of their geodesics locally coincides with classical hyperbolic geometry. They are important for the development of discrete differential geometry, since they were among the first where a meaningful discretization has been obtained [**Sau50, Wun51**], and because of the integrable equations associated with K-surfaces.

Like any other negatively curved surface in Euclidean 3-dimensional space, a K-surface has an asymptotic parametrization. It is not difficult to show that for a smooth asymptotic surface, the additional *Chebyshev condition*

$$\partial_2\|\partial_1 x\| = \partial_1\|\partial_2 x\| = 0 \tag{9}$$

characterizes $K = \text{const}$. Since the length of the derivative vectors w.r.t. one variable does not depend on the other variable, we can re-parametrize and achieve

$$\|\partial_1 x\| = \|\partial_2 x\| = 1. \tag{10}$$

It turns out that in this case, when traversing a parameter line with velocity 1, the surface's normal vector is rotating with velocity $\sqrt{-K}$. An obvious discretization of the Chebyshev property is

$$\Delta_2\|\Delta_1 x\| = \Delta_1\|\Delta_2 x\| = 0. \tag{11}$$

The construction of discrete K-surfaces depends on the properties of a so-called Chebyshev quad, where opposite edges have equal lengths

$$\|x - x_1\| = \|x_2 - x_{12}\| \quad \text{and} \quad \|x - x_2\| = \|x_1 - x_{12}\| \tag{12}$$

and which does not lie in a plane (*skew parallelogram*). The following statement is elementary spatial geometry:

LEMMA 3.1. *A non-planar quadrilateral x—x_1—x_{12}—x_2 with the Chebyshev property given by* (12) *(opposite edges have equal lengths) is symmetric w.r.t. a 180° rotation about the axis spanned by the two midpoints of diagonals.*

In each vertex x we consider a unit normal vector n orthogonal to the edges incident with x. The cyclic orientation of the quadrilateral face defines the orientation of the normal vector. The twist angle along an edge which is enclosed by the planes at either end (resp. by their normal vectors) is the same for opposite edges,

$$\langle n, n_1\rangle = \langle n_2, n_{12}\rangle\,, \quad \langle n, n_2\rangle = \langle n_1, n_{12}\rangle$$

(the unit normal vectors constitute a Chebyshev quad). Further, we can reconstruct, up to scaling, the edges of the quadrilateral from these normal vectors by

$$x_1 - x = n \times n_1, \qquad x_2 - x = -n \times n_2.$$

Also the converse is true: Every net of unit normal vectors which fulfills these conditions determines a K-net.

For a proof we refer to [**BS09**, §4.2.1]. The equal angles property is an immediate corollary of the symmetry. An illustration in [**Wun51**] shows a K-net which has actually been built, and where both the Chebyshev property and the angle property can be observed.

The systematic treatment of A-nets and K-nets $x(u)$ relies on the construction of the so-called Lelieuvre normal vector field $m(u)$, which has the property $x_k - x = m_k \times m$. It is a so-called T-net, the T-nets being an integrable 3-system. In this way the geometric considerations of [**Sau50, Wun51**] concerning skew-parallelogram lattices can be treated in a more high-level manner. We summarize, without proof:

THEOREM 3.2. *A discrete K-net $x(i_1, \dots, i_d)$ (an asymptotic net with the Chebyshev property) has unit normal vectors $n(i_1, \dots . i_d)$ which form a Chebyshev net themselves: The rotation angle of the normal vector along an edge xx_k depends only on the k-th parameter, and not on the others.*

The K-net property is a 2-system in the following sense: If unit normal vectors $n(i_1, 0)$ and $n(0, i_2)$ are given, then these data can be uniquely extended to a Chebyshev net $n(i_1, i_2)$ in the unit sphere, and a K-net x is derived from n by the formulae above.

The K-net property is d-dimensional consistent for $d \geq 3$ (i.e., integrable): Any choice of unit normal vectors

$$n(i_1, 0, \dots, 0), \quad n(0, i_1, \dots, 0), \quad \dots, \quad n(0, 0, \dots, i_d) \quad (d \geq 2)$$

can be extended to a Chebyshev net in the unit sphere, and define a K-net via the formulas given above.

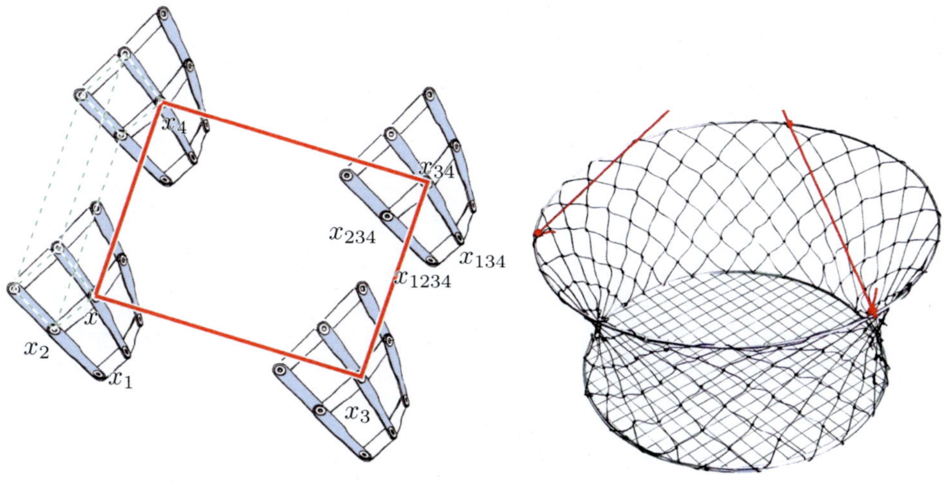

FIGURE 4. *Left:* A 4-dimensional discrete K-net, which at the same time is a 2D lattice of 2-dimensional K-nets, where each is in the Bäcklund transform relation with its neighbors. The thin dashed lines constitute a Bennett 12-bar linkage. *Right:* A K-net with rotational symmetry made from string (crab net).

3.3. Bäcklund transformation. Two smooth K-surfaces $x, x^{(1)}$, both parametrized such that parameter lines have unit speed, are Bäcklund transforms of each other, if the distance $\|x(u) - x^{(1)}(u)\|$ of corresponding points is constant, and if the line segment $x(u), x^{(1)}(u)$ is *tangent* to both surfaces in its endpoints. For a discrete surface $x(i_1, i_2)$ and its transform $x^{(1)}(i_1, i_2)$, the definition is literally the same, since even discrete K-nets possess tangent planes. Furthermore, it is clear from the definition that a K-net and its transform together constitute a 3-dimensional K-net.

Existence of 4D K-nets and the fact that they are determined by initial values (Theorem 3.2) shows Bianchi permutability of Bäcklund transforms, see Fig. 4.

3.4. Mechanisms based on skew parallelograms. The results on K-nets have as an immediate consequence the following remarkable fact which was first observed, at least in part, more than 100 years ago by G. T. Bennett [**Ben14**]:

Think of the K-net defined by normal vector data $n(i_1, 0)$ and $n(0, i_2)$. Deforming these initial values such that the angle between successive normal vectors remains constant yields a net $n(i_1, i_2)$ of normal vectors where the angle between neighbors remains constant. Computing the K-net from normal vector data yields a deformation of $x(i_1, i_2)$ such that both edge lengths and the twist along an edge remains constant.

Already a single quadrilateral is a nontrivial mechanism, called Bennett's four-bar mechanism or Bennett's isogram [**Ben14**]. An elementary cube in a 3-dimensional K-net is Bennett's 12-bar mechanism, see also Figure 4.

It is worth noting that certain K-nets can be manufactured rather easily, namely as equilibrium positions of fishing nets with rotational symmetry, see the crab net shown by Figure 4.

3.5. The sine-Gordon equation. By Gauss' *theorema egregium*, the Gaussian curvature K of a surface $x(u_1, u_2)$ can be computed from the coefficient functions $E = \langle \partial_1 x, \partial_1 x\rangle$, $F = \langle \partial_1 x, \partial_2 x\rangle$ and $G = \langle \partial_2 x, \partial_2 x\rangle$ of the first fundamental form. Francesco Brioschi's formula for Gaussian curvature says that

$$\begin{vmatrix} -\partial_{22}E + 2\partial_{12}F - \partial_{11}G & \partial_1 E & 2\partial_1 F - \partial_2 E \\ 2\partial_2 F - \partial_1 G & 2E & 2F \\ \partial_2 G & 2F & 2G \end{vmatrix} - \begin{vmatrix} 0 & \partial_2 E & \partial_1 G \\ \partial_2 E & 2E & 2F \\ \partial_1 G & 2F & 2G \end{vmatrix} = 8(EG - F^2)^2 K.$$

The unit speed (Chebyshev) property is expressed as $E = G = 1$, and with the angle $\phi(u_1, u_2)$ between parameter lines we have $F = \cos\phi$. Brioschi's formula yields $\partial_{12}\phi(u_1, u_2) = -K(u_1, u_2) \cdot \sin\phi(u_1, u_2)$. In the case of K-surfaces, it is no loss of generality to assume $K = -1$, so

$$\partial_{12}\phi = \sin\phi. \tag{13}$$

The meaning of this *sine-Gordon equation* equation in terms of the intrinsic geometry of a surface is this: If a surface is parametrized such that parameter lines are traversed with unit speed, and the angle between parameter lines evolves according to the sine-Gordon equation, then the surface has constant Gaussian curvature -1.

Note that Eqns. (10)+(13) constitute a purely intrinsic characterization of surfaces with $K = -1$, while Eqns. (10)+(7) provide an extrinsic characterization.

3.6. Discrete sine-Gordon equation. Integrable equations. A discrete K-surface where all edge lengths are equal to ε is the most direct analogue of a smooth surface parametrized by unit speed with Gaussian curvature -1. The evolution of the angle ϕ enclosed by edges xx_1 and xx_2 can be shown to obey the Hirota equation

$$\sin\frac{\phi_{12} - \phi_1 - \phi_2 + \phi}{4} = \frac{\varepsilon^2}{4}\sin\frac{\phi_{12} + \phi_1 + \phi_2 + \phi}{4}. \tag{14}$$

In these lecture notes we are not able to discuss the term *integrability* in a systematic manner, we refer to [**BS09**, §6] instead. We only mention some salient facts:

- integrable equations turn out to be closely related with multidimensionally consistent geometric properties of nets;
- discrete nets are easier to treat than continuous ones, in particular the special role of transformations disappears. The smooth case is obtained from the discrete case by a limit process;
- typical properties of integrable systems (Bäcklund transforms, zero curvature representations, ...) are a consequence of multidimensional consistency.

4. Applications: Computing minimal surfaces

There is a nice version of discrete minimal surfaces which connects an analytic approach with discrete geometry and circle patterns. It is also capable of solving a significant problem, namely computing the shape of a minimal surface from the combinatorics of its principal curve network [**BHS06**].

4.1. Isothermic surfaces and their duals. Isothermic surfaces represent a classical topic of differential geometry. Informally they are defined by the condition that their infinitesimal quadrilaterals are flat squares. Certain surface classes, including the minimal surfaces, admit an isothermic parametrization. It is a bit of a mystery why certain classes of surfaces have this property. For a good overview see [**Bur06**].

THEOREM 4.1. *An isothermic surface* $x : U \subseteq \mathbb{R}^2 \to \mathbb{R}^3$, *characterized by the condition that it is a conformal curvature-line parametrization, i.e.,*

$$\|\partial_1 x\| = \|\partial_2 x\| = s, \quad \langle \partial_1 x, \partial_2 x\rangle = 0, \quad \partial_{12}x \in \operatorname{span}(\partial_1 x, \partial_2 x),$$

with $s : U \to \mathbb{R}_+$, *has a Christoffel-dual surface* x^* *defined by*

$$\partial_1 x^* = \frac{1}{s^2}\partial_1 x \quad \partial_2 x^* = -\frac{1}{s^2}\partial_2 x. \tag{15}$$

x^* *is again isothermic, with* $\|\partial_j x^*\| = 1/s$.

PROOF. We have $\partial_{12}x = a\partial_1 x + b\partial_2 x$, for certain coefficient functions a, b. Further, $\partial_2(s^2) = 2\langle \partial_1 x, \partial_{12}x\rangle = 2\langle \partial_1 x, a\partial_1 x + b\partial_2 x\rangle = 2as^2$. Differentiating $\partial_1 x^*$ yields

$$\partial_2(\partial_1 x^*) = -\frac{\partial_2(s^2)}{s^4}\partial_1 x + \frac{1}{s^2}\partial_{12}x = -\frac{2a}{s^2}\partial_1 + \frac{1}{s^2}(a\partial_1 x + b\partial_2 x) = \frac{1}{s^2}(b\partial_2 x - a\partial_1 x).$$

For $\partial_1(\partial_2 x^*)$ an analogous computation yields the same result, so x^* exists. Obviously, x^* fulfills the isothermicity conditions. □

The Christoffel transformation is particularly interesting for *minimal surfaces*, since they admit isothermic parameters, and they are Christoffel-dual to their own unit normal vector field (which is then a conformal parametrization of the unit sphere). Conversely, any conformal parametrization of the unit sphere by Christoffel duality is converted into a minimal surface. If the unit sphere is identified with $\mathbb{C} \cup \{\infty\}$ by stereographic projection, and the definition of the Christoffel dual is written as an integral, this yields the Weierstrass representation of minimal surfaces.

Some parts of this relationship are easy to show. E.g. if x is a conformal parametrization of the unit sphere, then x is isothermic. The common unit normal vector field of x and the Christoffel dual x^* is furnished by x itself. Compute the principal curvatures of x^*:

$$\kappa_1 = \frac{\partial_1 x}{\partial_1 x^*} = s^2, \quad \kappa_2 = \frac{\partial_2 x}{bd_2 x^*} = -s^2.$$

i.e., x^* is a minimal surface. We also see how the speed of parametrization (which is s) is related to the values of the principal curvatures.

4.2. Discrete s-isothermic surfaces. Finding the "right" discretization of isothermic surfaces is an important topic in discrete differential geometry. The discretization shown here is one example of several – it is selected because of its success in solving a continuous problem by discretization. We start this discussion with a remarkable result on polyhedra. It can be seen as a discrete version of the Koebe uniformization theorem in the genus 0 case, and it can be made computationally efficient [**Sech07**].

THEOREM 4.2. *For each convex polyhedron there is a combinatorially equivalent convex "Koebe" polyhedron whose edges are tangent to the unit sphere. It becomes unique up to Euclidean transformations if we require that its center of mass equals* 0, *otherwise it is unique only up to Möbius transforms (i.e., up to projective automorphisms of the unit sphere).*

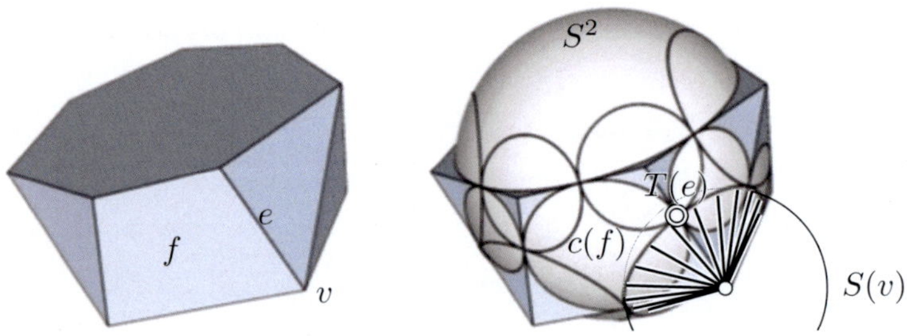

FIGURE 5. A Koebe polyhedron has edges tangent to a sphere. The orthogonal circle pattern defined by this polyhedron consists of the incircles of faces, and the circles of tangency of cones whose vertex is a vertex of the polyhedron. The vertex-centered sphere used in the definition of an s-isothermic surface is also shown.

Each Koebe polyhedron has two associated circle packings, see Fig. 5. System (i) arises by intersecting the faces' planes with the sphere; System (ii) arises in a dual way as the circles of tangency of cones whose vertex is a vertex of the polyhedron. System (ii) can also be replaced by vertex-centered spheres, showing that a Koebe polyhedron fulfills the definition of a discrete s-isothermic surface, see Fig. 6.

DEFINITION 4.3. A polyhedral surface (in particular a discrete conjugate net) is s-isothermic, if the following conditions are fulfilled.

- Every face f contains an incircle $c(f)$, every vertex v is the center of a sphere $S(v)$.
- Every edge e contains a point $T(e)$, such that:
  (*) $v \in e \implies S(v)$ intersects e orthogonally in $T(e)$
  (*) $e \subset f \implies c(f)$ touches e in $T(e)$
- Vertices have degree four, and faces have even degree.

FIGURE 6. An s-isothermic surface is defined as a circle/sphere arrangement where circles (green) touch circles, spheres (blue) touch spheres, and circles orthogonally intersect spheres. This example is a discrete Enneper surface (image courtesy T. Hoffmann).

FIGURE 7. Attempting to dualize a piece of Koebe polyhedron where one face has odd valence. The dual of the triangle does not close up (we cannot even assign labels ⊕,⊖ to edges in a consistent manner). However, if the given discrete surface is seen as a discrete branched covering, with the triangle actually being a hexagon, the dual surface will close after the original one is traversed twice (images courtesy B. Springborn).

If in addition the surface is part of a Koebe polyhedron, i.e., its edges are tangent to S^2, then we regard it as a discrete-conformal parametrization of S^2.

4.3. Dualizing s-isothermic surfaces. Consider a face $f = (v_0, \ldots, v_{n-1})$ of an s-isothermic surface with n even. Assume that the incircle of this face has radius ρ_f. The edge $v_i v_{i+1}$ touches the incircle in the point q_i. Then the dual face f^* likewise has an incircle of radius $\rho_{f^*} = 1/\rho_f$, and the new contact points q_i^* are defined by

(16)

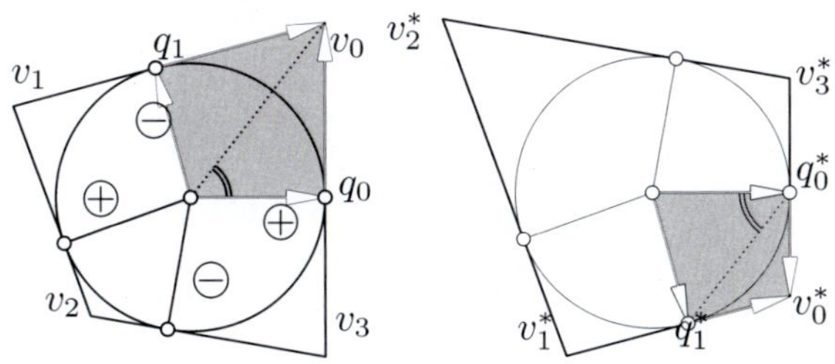

$$q_i^* = (-1)^i \frac{q_i}{\rho_f^2}$$

It follows that the edges of the dual and primal polygon are related by

$$p_j^* - p_{j+1}^* = (-1)^j \frac{1}{r_j r_{j+1}} (p_i - p_{i+1}), \text{ where } r_j = \|v_j - q_j\| = \|v_j - q_{j+1}\|. \tag{17}$$

The values r_j are the radii of the vertex-centered spheres which occur in the definition of an s-isothermic surface. The proof is an exercise in complex numbers. Similarly, one can show that the subdivision of the faces into quads as in (16) yields corresponding discrete surfaces $x(i_1, i_2)$ and $x^*(i_1, i_2)$ where

$$\Delta x_1^* = \frac{1}{\|\Delta x_1\|^2} \Delta x_1, \quad \Delta x_2^* = -\frac{1}{\|\Delta x_2\|^2} \Delta x_2. \tag{18}$$

This is completely analogous to the smooth case of (15).

For any given s-isothermic surface, we now assign labels ⊕ and ⊖ to all edges, such that for each face, the cycle of edges is assigned labels ⊕, ⊖, ⊕, ⊖, ... in an alternating way, indicating if the factor $+1$ or -1 is to be used in (16). By (17), the edge lengths of dual edges are independent of the face they are contained in. An interior angle $\alpha_{v,f}$ at the vertex v in the face f corresponds to the $\alpha^*_{v^*,f^*} = \pi - \alpha_{v,f}$ in the dual face, so the angle sum around the vertex v^* is

$$\sum_{f^* \ni v^*} \alpha^*_{v^*,f^*} = \sum_{f \ni v} (\pi - \alpha_{v,f}) = (\deg(v) - 2)\pi = 2\pi.$$

This shows that a discrete s-isothermic surface can be dualized.

DEFINITION 4.4. A discrete s-isothermic surface is minimal, if its dual is spherical, i.e., is part of a Koebe polyhedron.

[**BHS06**] discusses how to compute the shape of a minimal surface from the combinatorics of its network of principal curvatures lines. One starts by computing a Koebe polyhedron with the desired combinatorics, and dualizes, see Fig. 7. One can even show convergence to smooth minimal surface, by using the fact that correspondences between combinatorially equivalent circle packings can be shown to converge to conformal mappings [**HS93**].

4.4. Curvatures of discrete surfaces. Discrete surfaces have been assigned curvatures in different ways. E.g. the variational definition of the mean curvature vector field $\vec{H}$ as the gradient of the area functional leads to *cotangent formula* for mean curvature of simplicial surfaces. Another approach uses the *Steiner formula*: A surface in $\mathbb{R}^3$, with unit normal vector n, has constant-distance offset surfaces where a point p of the original surface moves to $p + tn(p)$. Surface area changes according to

$$A^t = \int (1 - 2Ht + Kt^2)\, dA, \tag{19}$$

where dA refers to the canonical surface area measure, and H, K are mean curvature and Gaussian curvature, respectively.

[**PLW⁺07, BPW10**] proposed to use the same principle for a polyhedral surface (V, E, F) equipped with a polyhedral Gauss image $(\sigma(V), \sigma(E), \sigma(F))$ (i.e., normal vector field), such that corresponding faces of surface and Gauss image are parallel. An example is furnished by the s-isothermic minimal surfaces and the corresponding Koebe polyhedron. We define a constant-distance offset (V^t, E^t, F^t) by vertices

$$v^t = v + t\sigma(v).$$

The oriented area of closed polygon $f = (v_0, \dots, v_{n-1})$ in $\mathbb{R}^2$ is a quadratic form $A(f)$ with associated symmetric bilinear form $A(f, g)$,

$$A(f) = \frac{1}{2} \sum\nolimits_{i=0}^{n-1} \det(v_i, v_{i+1}), \quad A(f, g) = \frac{1}{2}(A(f + g) - A(f) - A(g))$$

(Leibniz' sector formula). $A(f, g)$ is called the oriented mixed area of f, g. Then the area of a face $f^t \in F^t$ is given by

$$A(f^t) = A(f) + 2tA(f, \sigma(f)) + t^2 A(\sigma(f)). \tag{20}$$

Comparing (19) with (20) leads to the definition of a mean curvature H and a Gaussian curvature K of a face f:

$$H(f) = -\frac{A(f, \sigma(f))}{A(f)}, \quad K(f) = -\frac{A(\sigma(f)}{A(f)}. \tag{21}$$

With this definition it is not difficult to compute the mean curvature of s-isothermic surfaces: it turns out to be zero. The considerations above have been extended to

FIGURE 8. This torsion-free support structure (image courtesy Evolute GmbH) guides members and nodes in the outer hull of the Yas Marina hotel in Abu Dhabi, so that members have a nice intersection in each node (at right, image courtesy Waagner-Biro Steel and Glass). Note that the faces of this mesh are not planar.

more general surface classes by [**HSFW17**]. E.g the discrete K-surfaces indeed enjoy the property that the Gaussian curvature equals -1.

5. Applications: Freeform architecture

With computer-aided design, it has become rather easy for professionals to design free forms in architecture, but building them is another matter. The geometric questions which arise in this context have a rich connection to discrete differential geometry. Recent surveys are [**PW16**, **PEVW15**]. Examples of specific geometric issues in the context of freeform architecture are the following.

• Steel-glass constructions often are discrete surfaces with flat faces, because of the glass panels that are put there. For surfaces with quadrilateral faces, the planarity requirement is a nontrivial side-condition. Another issue with steel constructions is the manner of intersection of the beams in nodes. We discuss this topic below, see also [**LPW$^+$06**, **PLW$^+$07**].

• Freeform skins might be required to be at constant distance from each other, or a steel-glass construction might be required to be of constant thickness. This topic leads to circular nets, conical nets, and even nets which are edgewise parallel to a Koebe polyhedron, depending on the question how distances are measured (between vertices, or faces, or edges), see [**PLW$^+$07**].

• Developable surfaces occur in bent glass and in curved beams (whose sides are made by bending flat pieces of steel). A sequence of developables is a semidiscrete conjugate net, a viewpoint which has been exploited by [**LPW$^+$06**, **PSB$^+$08**].

• The differential geometry of manifolds in line space, and discrete submanifolds of line space has been used by [**WJB$^+$13**] to study lighting and shading.

• Circle packings and discrete uniformization turn up in the question of regular patterns, in particular hexagonal patterns [**SHWP09**].

• Self-supporting surfaces like brick vaults and their polyhedral Airy potential surface are related to isotropic geometry and curvatures [**VHWP12**]. A discrete stress state and corresponding polyhedral Airy potential is also relevant for material-minimizing structures [**PKWP17**].

5.1. Meshes. Torsion-free support structures. Here we discuss briefly some topics from the list which have to do with discrete *parametric* surfaces. In order to describe structures from straight or curved beams, with or without panels covering the faces, we start with the combinatorics. A mesh (V, E, F) consists of a graph (V, E), where individual edge cycles are designated as faces $f \in F$. This has to be done in such a way that the complex glued from points (corresponding to vertices), line segments (corresponding to edges) and disks (one for each face) is a topological manifold. In a geometric realization of the connectivity, edges are realized as line segments or as curves. Faces are filled by planar or non-planar surfaces.

Many freeform architectural structures can be modelled by means of meshes, see e.g. Figure 8 for a mesh where faces are not filled in, but edges are represented by straight beams and vertices by connectors between the individual beams.

If faces are combinatorial triangles, resp. quads or hexagons, we speak of a triangle mesh, resp. quad mesh or hexagonal mesh. Terminology is often sloppy and one applies the term quad mesh or hex mesh also in cases where most faces are quads, or most are hexagons. The discrete parametric surfaces mapping from $\mathbb{Z}^2$ to $\mathbb{R}^3$ are quad meshes in a natural way.

Definition 5.1. A torsion-free support structure associated to a mesh (V, E, F) with straight edges is an assignment of a line $\ell(v)$ to each vertex $v \in V$ and a plane $\pi(e)$ to each edge $e \in E$, such that $v \in e \implies \ell(v) \subset \pi(e)$ (plus certain nondegeneracy conditions).

If the edges of the mesh are curves, $\pi(e)$ is a developable surface containing the edge.

The relevance of a torsion-free support structure is that straight beams in a steel construction can be aligned with the planes $\pi(e)$, and the symmetry planes of beams nicely intersect in the line $\ell(v)$, whenever beams come together in a node, see Fig. 8. Such an intersection of beams is called torsion-free. If the intersection is not torsion-free, and beams intersect anyhow, the intersection is complicated to manufacture, see Fig. 9. However, if the intersection is torsion-free, one can simply use a cylindrical node element and connect the beams to it.

Proposition 5.2. *A properly connected triangle mesh has only trivial support structures, where all elements pass through a single center z: for all $v \in V, e \in E$ we have $\ell(v) = v \vee z$ and $\pi(e) = e \vee z$.*

Proof. In projective space, and ignoring degeneracies, we argue as follows. In a face $f = v_i v_j v_k$, we have $\ell(v_i) = \pi(v_i v_j) \cap \pi(v_i v_k)$, and similar for $\ell(v_j)$, $\ell(v_k)$. Thus, the point $z(f) = \pi(v_i v_j) \cap \pi(v_i v_k) \cap \pi(v_j v_k)$ lies on all lines $\ell(v_i)$, $\ell(v_j)$, $\ell(v_k)$. If $f' = (v_i v_j v_l)$ is a neighbor face, then $z(f) = \ell(v_i) \cap \ell(v_j) = z(f')$. By connectedness, $z(f)$ is the same point for all faces $f \in F$. □

This result is one reason why quad meshes are attractive for steel-glass constructions. Unfortunately, imposing planarity of faces on quad meshes is nontrivial.

5.2. The design dilemma. When using mathematical methods in an artistic context, one faces a problem which is not known from applications in the natural sciences: It is usually not desirable that Mathematics yields a definite answer to a certain question. We illustrate this by means of torsion-free support structures: Such a structure either consists of developable surfaces, or of discrete developables following the edges of the mesh. If they are to be *orthogonal* to the reference surface, like in the case of Figure 9, left, then they must follow the surface's principal

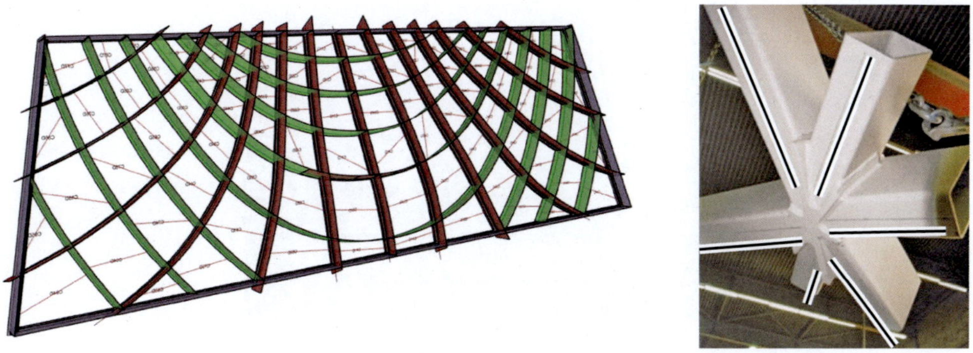

FIGURE 9. *Left:* A curved support structure (Eiffel tower pavilions, Paris. Image Evolute GmbH, 2010). *Right:* An intersection of beams which does not follow a torsion-free support structure (courtesy Waagner-Biro Steel and Glass).

curvature lines, cf. the discussion at the end of §2. Since the principal curvature lines are determined by the reference surface, there is no longer any design freedom for the beams except perhaps their spacing. Such a situation, which amounts to a restriction of the artist's freedom of expression, is unacceptable to the designer. In the case of the Eiffel tower pavilions, design freedom was restored by the fact that small changes to the design surface can cause big changes in the network of principal curvature lines. In this way it was possible to achieve the desired layout of beams by only minimally changing the reference surface.

5.3. Meshes with planar faces – conjugate nets. The design of quadrilateral meshes with planar faces was among the first topics where a connection between freeform architecture and discrete differential geometry was established [**LPW$^+$06**]. If the mesh under consideration has the visual appearance of a smooth network of curves on a smooth surface, then such a quad mesh has mostly regular combinatorics and is therefore a conjugate net which approximates a conjugate curve network, see Fig. 2. If in addition to conjugacy we add orthogonality of curves, we have the principal network. In fact, any conjugate network useful for deriving a quad mesh from it is close to principal, which is another instance of the design dilemma mentioned above. Since the reference shape and a conjugate net on it cannot be designed separately, there are very few instances of such meshes which serve as basis of actual building skins, and most of those have been generated by a simple process like translating one polygon along another one, see Fig. 10.

5.4. Constant-distance offsets. In the context of meshes with planar faces, a natural question is the existence of another mesh which is combinatorially equivalent and which is *at constant distance* from the first one. One can consider face-face distances, or edge-edge distances, or vertex-vertex distancesconstant. In order to treat these cases in an uniform manner, we first define parallelity of meshes:

DEFINITION 5.3. Consider combinatorially equivalent meshes (V, E, F) and (V', E', F') with straight edges and planar faces. They are parallel, if corresponding edges of E resp. E' lie in parallel lines, and corresponding faces of F resp. F' lie in parallel planes (details regarding nondegeneracy have been omitted).

FIGURE 10. This quad mesh with parallelogram faces has been constructed by parallel-translating a polygon along another polygon (Hippo house, Berlin Zoo. Engineering by Schlaich Bergermann & Partners).

A conjugate net and its parallel net together constitute a 3D conjugate net. It is not difficult to show the following [**PLW+07**]:

• If (V', E', F') is parallel to (V, E, F), then a torsion-free support structure is defined by $\ell(v) = v \vee v'$ and $\pi(e) = e \vee e'$, where v, v' and e, e' are pairs of corresponding vertices resp. edges. In the simply connected case, a torsion-free support structure also implies existence of parallel meshes

• Parallel meshes which are at constant vertex-vertex distance d in a nontrivial way are circular. This is because we can construct a third mesh by computing vertices according to $v'' = (v' - v)/d$, which is inscribed to the unit sphere and has planar faces, so it is circular. For a quadrilateral, circularity is expressed via the angles between edges only, so the two meshes we start with inherit the circular property.

These two statements show that conjugate nets and circular nets we studied in the beginning, have actual applications in a field which generally was not known for using much applied Mathematics at all, namely the design of freeform skins for architecture.

References

[Bäck83] Albert Bäcklund, *Om ytor med konstant negativ krökning*, Lunds Univ. Årsskr., vol. 19, 1883, 42 pages.

[Ben14] G. T. Bennett, *The Skew Isogram Mechanism*, Proc. London Math. Soc. (2) **13** (1914), 151–173, DOI 10.1112/plms/s2-13.1.151. MR1577497

[BHS06] Alexander I. Bobenko, Tim Hoffmann, and Boris A. Springborn, *Minimal surfaces from circle patterns: geometry from combinatorics*, Ann. of Math. (2) **164** (2006), no. 1, 231–264, DOI 10.4007/annals.2006.164.231. MR2233848

[Bia02] Luigi Bianchi, *Lezioni di geometria differenziale*, 2nd ed., Spoerri, Pisa, 1902.

[BP96] Alexander Bobenko and Ulrich Pinkall, *Discrete surfaces with constant negative Gaussian curvature and the Hirota equation*, J. Differential Geom. **43** (1996), no. 3, 527–611. MR1412677

[BPW10] Alexander I. Bobenko, Helmut Pottmann, and Johannes Wallner, *A curvature theory for discrete surfaces based on mesh parallelity*, Math. Ann. **348** (2010), no. 1, 1–24, DOI 10.1007/s00208-009-0467-9. MR2657431

[BS09] Alexander I. Bobenko and Yuri B. Suris, *Discrete differential geometry*, Graduate Studies in Mathematics, vol. 98, American Mathematical Society, Providence, RI, 2008. Integrable structure. MR2467378

[Bur06] F. E. Burstall, *Isothermic surfaces: conformal geometry, Clifford algebras and integrable systems*, Integrable systems, geometry, and topology, AMS/IP Stud. Adv. Math., vol. 36, Amer. Math. Soc., Providence, RI, 2006, pp. 1–82. MR2222512

[Eis23] Luther P. Eisenhart, *Transformations of surfaces*, Princeton University Press, 1923.

[HS93] Zheng-Xu He and Oded Schramm, *Fixed points, Koebe uniformization and circle packings*, Ann. of Math. (2) **137** (1993), no. 2, 369–406, DOI 10.2307/2946541. MR1207210

[HSFW17] Tim Hoffmann, Andrew O. Sageman-Furnas, and Max Wardetzky, *A discrete parametrized surface theory in* $\mathbb{R}^3$, Int. Math. Res. Not. IMRN **14** (2017), 4217–4258, DOI 10.1093/imrn/rnw015. MR3674170

[LPW+06] Yang Liu, Helmut Pottmann, Johannes Wallner, Yong-Liang Yang, and Wenping Wang, *Geometric modeling with conical meshes and developable surfaces*, ACM Trans. Graph. **25** (2006), no. 3, 681–689.

[PEVW15] Helmut Pottmann, Michael Eigensatz, Amir Vaxman, and Johannes Wallner, *Architectural geometry*, Computers and Graphics **47** (2015), 145–164.

[PKWP17] Davide Pellis, Martin Kilian, Johannes Wallner, and Helmut Pottmann, *Material-minimizing forms and structures*, ACM Trans. Graphics **36** (2017), no. 6, article 173.

[PLW+07] Helmut Pottmann, Yang Liu, Johannes Wallner, Alexander Bobenko, and Wenping Wang, *Geometry of multi-layer freeform structures for architecture*, ACM Trans. Graph. **26** (2007), article 65.

[PP93] Ulrich Pinkall and Konrad Polthier, *Computing discrete minimal surfaces and their conjugates*, Experiment. Math. **2** (1993), no. 1, 15–36. MR1246481

[PSB+08] Helmut Pottmann, Alexander Schiftner, Pengbo Bo, Heinz Schmiedhofer, Wenping Wang, Niccolo Baldassini, and Johannes Wallner, *Freeform surfaces from single curved panels*, ACM Trans. Graph. **27** (2008), no. 3, article 76.

[PW16] Helmut Pottmann and Johannes Wallner, *Geometry and freeform architecture*, Mathematics and society, Eur. Math. Soc., Zürich, 2016, pp. 131–151. MR3497592

[Sau50] Robert Sauer, *Parallelogrammgitter als Modelle pseudosphärischer Flächen* (German), Math. Z. **52** (1950), 611–622, DOI 10.1007/BF02230715. MR37042

[Sech07] Stefan Sechelmann, *Discrete minimal surfaces, Koebe polyhedra, and Alexandrov's theorem. variational principles, algorithms and implementation*, Master's thesis, TU Berlin, 2007.

[SHWP09] Alexander Schiftner, Mathias Höbinger, Johannes Wallner, and Helmut Pottmann, *Packing circles and spheres on surfaces*, ACM Trans. Graph. **28** (2009), no. 5, article 139.

[VHWP12] Etienne Vouga, Mathias Höbinger, Johannes Wallner, and Helmut Pottmann, *Design of self-supporting surfaces*, ACM Trans. Graph. **31** (2012), no. 4, article 87.

[WJB+13] Jun Wang, Caigui Jiang, Philippe Bompas, Johannes Wallner, and Helmut Pottmann, *Discrete line congruences for shading and lighting*, Comput. Graph. Forum **32** (2013), no. 5, 53–62.

[Wun51] Walter Wunderlich, *Zur Differenzengeometrie der Flächen konstanter negativer Krümmung* (German), Österreich. Akad. Wiss. Math.-Nat. Kl. S.-B. IIa **160** (1951), 39–77. MR56342

TU Graz, 5070 Institut für Geometrie, 8010 Graz, Kopernikusgasse 24/IV
Email address: j.wallner@tugraz.at

Proceedings of Symposia in Applied Mathematics
Volume **76**, 2020
https://doi.org/10.1090/psapm/076/00671

Discrete mappings

Yaron Lipman

ABSTRACT. Piecewise-linear triangle meshes are widely popular for surface representation in the digital computer; mappings of meshes are therefore central for applications such as computing good coordinate systems on surfaces (parameterization), finding correspondence between shapes, and physical simulation. However, representation and calculation of mappings in a computer pose several challenges: (i) how to define faithful discrete analogous of properties of smooth mappings (e.g., angle or area preservation), (ii) how to guarantee properties such as injectivity and/or surjectivity, and (iii) how to construct mappings between non-Euclidean (curved) domains. One particularly interesting sub-class of simplicial mappings is the collection of convex combination mappings, in which the image of each vertex of the triangulation is restricted to the convex-hull of its immediate neighbors' images. Convex combination mappings can guarantee injectivity, are simple to compute algorithmically, offer a discrete analog to harmonic mappings, and can be used to approximate conformal mappings. This lecture provides an introduction to convex combination mappings and their generalizations, as well as their algorithmic aspects and practical applications.

1. Triangulations and discrete mappings

Surfaces in a computer are often represented or approximated by triangulations $M = (V, E, F)$, where $V = \{v_i\} \subset \mathbb{R}^d$ (typically $d = 2, 3$) is the vertex set, $E = \{e_{ij}\}$ the edge set, and $F = \{f_{ijk}\}$ the face set. Edges are convex-hulls of pairs of points, $e_{ij} = \text{hull}\{v_i, v_j\}$, and faces are convex-hulls of triplets of points $f_{ijk} = \text{hull}\{v_i, v_j, v_k\}$. M is a simplicial complex, meaning that:

(i) The intersection of any two faces is either an edge, a vertex or an empty set; the intersection of any two edges is either a vertex or empty.
(ii) All edges of a triangle are in E; all vertices of an edge are in V.

Note that a triangulation M is not necessarily a topological surface (i.e., each point has a neighborhood homeomorphic to a disk), as shown for example in Figure 1(a). To assure a triangulation is indeed a topological surface we add the requirement that

(iii) The link of each vertex is a simple closed polygon.

2020 *Mathematics Subject Classification.* Primary 57R35; Secondary 58K20.
Key words and phrases. Discrete differential geometry, integrable systems.
This work was funded by the Israel Science Foundation (grant No. 1830/17).

The link of a vertex v_i is the union of all edges e_{jk} that do not contain v_i but share a triangle f_{ijk} with it, namely

$$\mathrm{link}(v_i) = \bigcup_{\{e_{jk} \mid f_{ijk} \in F\}} e_{jk}.$$

DEFINITION 1. A *surface triangulation* is a triangulation $M = (V, E, F)$ satisfying (i)-(iii).

For example, Figure 1(a)-left is not a surface triangulation since the link of the middle vertex consists of two closed polygonal loop; the right example is not a surface since the link of a vertex at any of the two ends of the center edge contains a subset of edges homeomorphic to the symbol Y.

To allow for surface triangulations with boundary we relax Condition (iii):

(iii') The link of each vertex is a simple closed polygon or a simple polygonal arc.

DEFINITION 2. A *triangulation of a surface with a boundary* is a triangulation $M = (V, E, F)$ satisfying (i)-(ii) and (iii').

The boundary of a surface triangulation M is a one-dimensional simplicial complex, that is a polygonal curve. It consists of all edges with only one adjacent face and all vertices whose link is a polygonal arc. We denote it by $M_B = (V_B, E_B)$. The interior vertex set is denoted $V_I = V \setminus V_B$. Note that in general, in contrast to our intuition from the continuous case, not every boundary vertex has an interior vertex as a neighbor; e.g., stitch a triangle to the boundary of a disk-type mesh and consider its dangling vertex.

A surface triangulation $M = (V, E, F)$ is connected if its underlying graph $V = (V, E)$ is connected. It is 3-connected if it cannot be disconnected by removing any two vertices. If M is a boundaryless surface triangulation it is automatically 3-connected. If M is disk-type and 3-connected the relation of boundary vertices V_B to interior vertices B_I is similar to the continuous case: every boundary vertex has an interior vertex neighbor (if interior vertices exist). In fact, a stronger claim holds in the 3-connected case [**Flo03a**]: any interior vertex v_i can be connected to any other vertex v by a path $P = [v_i, v_1, v_2, \ldots, v_k, v]$ of interior vertices $v_j \in V_I$, $1 \le j \le k$.

We want to discuss mappings of surface triangulations. The most natural class of mappings of a triangulation consists of simplicial maps:

DEFINITION 3. A *simplicial map* $f : M \to \mathbb{R}^d$ is a continuous piecewise-linear mapping defined as the unique piecewise-linear extension of a vertex map taking vertices $v_i \in V$ to $u_i \in \mathbb{R}^d$. By linear extension we mean that every point in the interior of some simplex σ (edge or face) $x = \sum_i \lambda_i v_i$, where $\lambda_i \ge 0$, $\sum_i \lambda_i = 1$, and v_i are the vertices of σ, is mapped to $f(x) = \sum_i \lambda_i u_i$.

A guiding problem for us is:

PROBLEM 1. Given two topologically equivalent surfaces[1] M_1, M_2, compute a simplicial homeomorphism $f : M_1 \to M_2$. We would like this homeomorphism to have some minimal distortion property and potentially interpolate given point data $\{(x_i, y_i)\}_{i \in I} \subset M_1 \times M_2$.

[1]Two surfaces are topologically equivalent if there exists a homeomorphism between them. Intuitively, the surfaces can be stretched onto one another without the need to tear the surfaces.

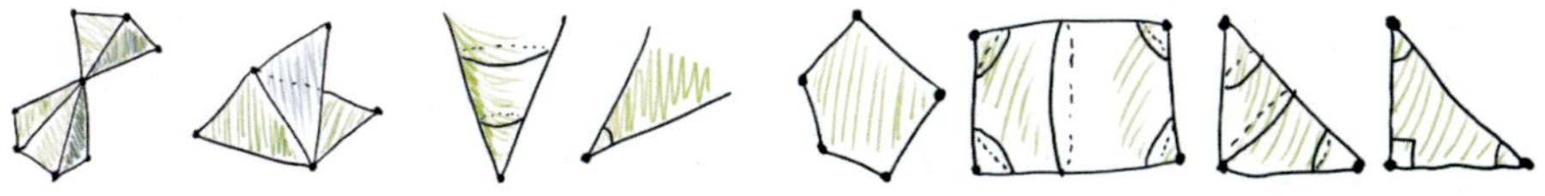

FIGURE 1. (a) Non manifold triangulations; (b) a cone (left) and a sector (right); (c) Euclidean cone surfaces: convex polygonal region (left), Euclidean orbifolds (middle, right).

Unfortunately, directly computing a homeomorphic simplicial map $M_1 \to M_2$ is in general a non-convex and difficult problem. We will tackle this problem by considering a canonical (topologically equivalent) domain $\mathcal{N}$ and show how to compute simplicial homeomorphisms $f_1 : M_1 \to \mathcal{N}$ and $f_2 : M_2 \to \mathcal{N}$. We then define $f = f_2^{-1} \circ f_1 : M_1 \to M_2$. The canonical surface $\mathcal{N}$ will be chosen from a particular family of special surfaces $\mathcal{F} = \{\mathcal{N}\}$ we will consider in this chapter.

2. Convex combination mappings

Our goal is to build a simplicial homeomorphism $M \to \mathcal{N}$ onto a member surface of a special family of surfaces $\mathcal{F} = \{\mathcal{N}\}$. Building the homeomorphism $M \to \mathcal{N}$ will be accomplished by solving a system of sparse linear equations. The method of constructing this simplicial map is called *convex combination mapping.* It originated in [**Tut63**] and was generalized and given this name in [**Flo97**, **Flo03a**, **RG06**]. These works allowed homeomorphic simplicial mappings of topological disks onto convex polygonal regions. In [**Lov04**, **SF04**, **GGT06**] the embedding was generalized to the topological torus. In [**GGT06**] sufficient conditions for non-convex, as well as multiply connected domains were formulated. In [**AL15**], the construction was generalized to the collection of Euclidean orbifolds (which contains the torus as a particular case). The following is based mostly on the above papers.

$\mathcal{F}$ will soon be defined as a certain collection of Euclidean cone surfaces.

DEFINITION 4. A compact oriented surface $\mathcal{N}$ is a *Euclidean cone surface* if it is a metric space locally isometric to an open disk, a cone, or a sector, and the number of cone points is finite.

A cone and a sector are shown in Figure 1(b). For each point $x \in \mathcal{N}$ we define the angle $\theta(x)$ to be the angle sum at the point x. Interior points have $\theta(x) = 2\pi$ and boundary points $\theta(x) = \pi$. An interior point is a cone point if $\theta(x) \neq 2\pi$. A boundary point is a cone point if $\theta(x) \neq \pi$. We will consider a specific family of Euclidean cone surfaces:

$$\mathcal{F} = \{\mathcal{N} \mid \mathcal{N} \text{ is a convex polygonal domain or a Euclidean orbifold}\}.$$

A convex polygonal domain is shown in Figure 1(c), left. Figure 1(c), right, shows three examples of Euclidean orbifolds: two topological spheres and a topological disk. Euclidean orbifolds is a specific family of Euclidean cone surfaces defined by taking the quotient of the plane $\mathbb{R}^2$ w.r.t. a wallpaper symmetry group of the plane, G, i.e., $\mathbb{R}^2/G$. Put differently, a symmetry group G defines an equivalence relation $u \sim w$ iff $u = g(w)$ for some $g \in G$. Taking the quotient topology w.r.t. such an equivalence relation leads to a surface $\mathcal{N}$ where the orbits of G, namely

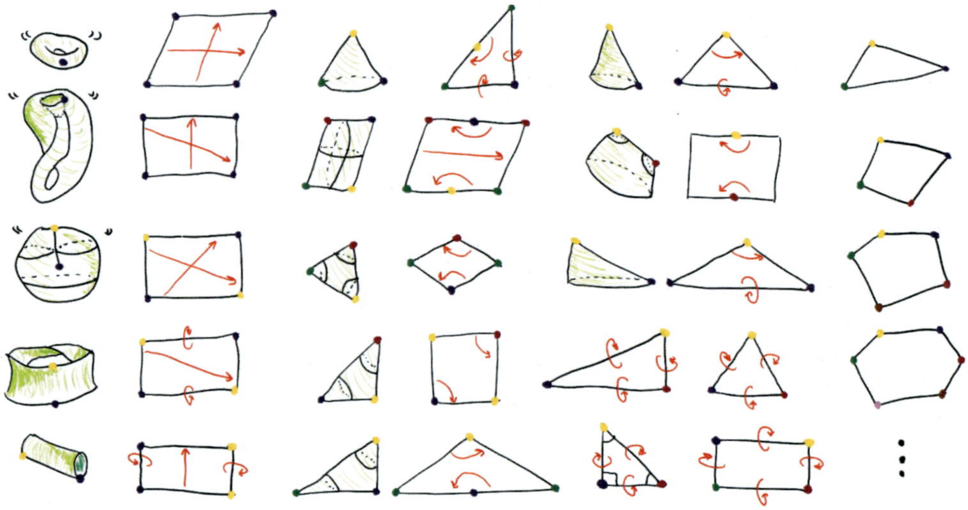

FIGURE 2. All types of fundamental domains; rigid motion identifications of edges are depicted with arrows.

$[u] = \{g(u) \mid g \in G\}$, are identified as points. The orbifolds are in one-to-one correspondence with the 17 wallpaper groups. Euclidean orbifolds are characterized by their topological type, and number, order and types of cones. There are two types of cones: reflective (where the surface is locally isometric to a sector, see Figure 1(b), right), and rotational (where the surface is locally isometric to a cone, see Figure 1(b), left). The order of a cone point x is $2\pi/\theta(x)$ for rotational and $\pi/\theta(x)$ for reflectional cones. Figure 3 lists all Euclidean orbifolds using the so-called orbifold notation [**CBGS08**]. Figure 1(c) shows (2222), (244), and (*244) orbifolds, respectively.

Each Euclidean orbifold has a (non-unique) fundamental domain, namely a domain that represents a connected choice of a representative per orbit. As fundamental domains we use (closed) disk-type polygonal domains Ω defined by a closed convex polygonal curve $[p_1, p_2, \ldots, p_m]$ together with a list of identifications of pairs of edges $[p_i, p_{i+1}] \leftrightarrow [p_j, p_{j+1}]$ by rigid motions $g \in G$, i.e., $g([p_i, p_{i+1}]) = [p_j, p_{j+1}]$. Note that these rigid identifications g also generate G. Pairs of different edges are identified with rotations, translations, or glides (a composition of a reflection and translation along a line), while self identified edges are identified with reflections. Figure 2 shows all fundamental domains where the identification are visualized by arrows. Arrows between different edges include rotations, translations, and glides (slanted arrows); self arrows indicate reflection across the infinite line supporting the edge. The edge identification implies an equivalence relation in $\Omega \times \Omega$ where identified points are in the same equivalence class. Once we quotient Ω by this relation (i.e., the edges of Ω are stitched together) we achieve the orbifold $\mathcal{N}$. For convex polygonal domains we denote $\Omega = \mathcal{N}$ with all edges self-identified (right column in Figure 2).

We will build a simplicial mapping $f : M \to \mathcal{N}$ called a convex combination mapping. This is done by assigning a positive weight, $\omega_{ij} > 0$, to each edge $e_{ij} \in E$.

topology	cones
sphere	(236), (244), (333), (2222)
disk	$(*236)$, $(*244)$, $(*333)$, $(*2222)$, $(2*22)$, $(3*3)$, $(4*2)$, $(22*)$
projective plane	$(22x)$
torus	(o)
Klein bottle	(xx)
annulus	$(**)$
Möbius band	$(*x)$

FIGURE 3. The types of Euclidean orbifolds. On left to $*$ (or without any $*$) are rotational cone orders; on the right to $*$ reflective cone orders.

DEFINITION 5. A *convex combination mapping* $f : M \to \mathbb{R}^2$ is a simplicial map, mapping each interior vertex $v_i \in V_I$ to a planar point $u_i \in \mathbb{R}^2$ so that

$$\sum_{j \in \mathfrak{N}_i} \omega_{ij}(u_j - u_i) = 0, \tag{1}$$

where $\mathfrak{N}_i = \{j \mid e_{ij} \in E\}$ is the neighbor index set of v_i. Geometrically, u_i is strictly inside the convex-hull defined by its immediate neighbors.

An important property of convex combination mappings is the discrete maximum principle. It is useful to formulate it in the functional setting: Let $h : M \to \mathbb{R}$ satisfy the convex combination property (1) at each interior vertex V_I. That is, to each vertex $v_i \in M$ one associates a scalar $h_i \in \mathbb{R}$ and these scalars satisfy (1) for all $v_i \in V_I$. For example, one can consider the x- or y-coordinate of convex combination mapping. Then, as proved e.g., in [**Flo03a**],

PROPOSITION 1. *(Discrete maximum principle.) Let $h : M \to \mathbb{R}$ be a convex combination function, and M a 3-connected disk-type surface triangulation. Let $v_i \in V_I$. If $h_i = \min_j h_j$ or $h_i = \max_j h_j$ then h is a constant function. In particular this implies that the maximum and minimum is achieved on the boundary V_B. The last assertion is true also if M is not 3-connected.*

PROOF. Assume $h_i = \min h_j$. The convex combination property together with the fact that h_i achieves the minimum imply that its immediate neighbors also equal h_i. Using the 3-connectedness we can construct an interior path to any other vertex, therefore continuing in this manner the proposition is proved. □

Convex combination mappings by themselves are not sufficient for building homeomorphisms. For example, the trivial (i.e., constant) convex combination mapping $u_i = u$, $\forall i$ always exists. To achieve a homemorphism certain boundary conditions should be applied. These boundary conditions in essence force the image of the map, $f(M)$, to cover $\mathcal{N}$. Bijectivity follows from the particular properties of the family of surfaces $\mathcal{F}$ and is not true in general for all Euclidean cone surfaces.

We deal with M that has one of the topological types that appear in $\mathcal{F}$ (all possible topological types and fundamental domains are listed in Figure 3 and depicted in Figure 2). Given a choice of target cone surface $\mathcal{N} \in \mathcal{F}$ with m cone points, let its fundamental disk domain be denoted by Ω. Choose a simple connected polygonal path $\Gamma \subset V \cup E$ passing through at-least m vertices of M, denoted $C = \{v_{c_1}, v_{c_2}, \ldots, v_{c_m}\} \subset V$, so that $M \setminus \Gamma$ is homeomorphic to Ω° (i.e., the interior

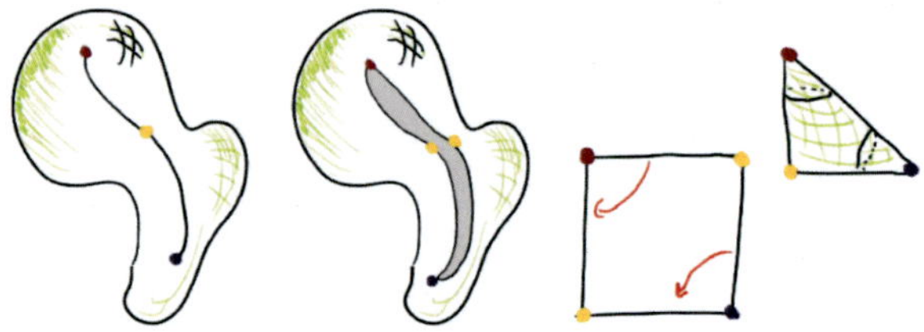

FIGURE 4. A surface triangulation M (left) is cut into M' (middle-left) to be homeomorphic to the fundamental domain Ω (middle-right) of a selected target $\mathcal{N}$ (right).

of Ω). Let $M' = \{V', E', F'\}$ be the triangulation representing M cut along Γ, where $F' = F$ and M' is homeomorphic to Ω. Figure 4 shows an example of (from left to right): triangulations M, M', Ω and a target orbifold $\mathcal{N}$.

We will need to relate the vertices, edges, and faces of M' and M. For that end we consider the inclusion map $\iota : M' \hookrightarrow M$. The inclusion ι induces an equivalence relation in M': $x \sim y \iff \iota(x) = \iota(y)$. Taking the quotient of M' with this equivalence relation results in M; this is equivalent to stitching M' along Γ to retrieve back M.

To compute the simplicial homeomorphism $f : M \to \mathcal{N}$ we consider M' and define a simplicial map $s : M' \to \mathbb{R}^2$ by solving a linear system of equations. The unknowns of the linear system are the target locations of the vertices in the plane, namely $u_i = s(v_i)$, $v_i \in V'$; the simplicial map s is the unique piecewise linear extension of this vertex map. The total degrees of freedom of s are therefore $u = \{u_1, u_2, \ldots, u_{|V'|}\} \in \mathbb{R}^{2\times|V'|}$. The linear system consists of three sets of equations:

(a) For interior vertices, V'_I, we set the convex combination equation (1), that is, every interior vertex is mapped strictly inside the convex hull of its immediate neighbors.

(b) For (non-cone) boundary vertices, $V'_B \setminus C'$, where $C' = \iota^{-1}(C) \subset V'$, we formulate equations preserving the edge identification rigid motions (see Figure 2), and maintaining the convex combination condition across the boundary of M'; Equations (3),(4),(5),(6).

(c) For cone vertices C', we restrict their position to the vertices of the fundamental domain Ω,

$$u_i = p_i, \quad v_i \in C'. \tag{2}$$

Equations (b) are dependent on the choice of target surface $\mathcal{N} \in \mathcal{F}$ and realize the edge identification in the fundamental domain. The equivalence relation $\sim$ induces an equivalence relation in the vertex set, $v_i \sim v_j$ in V'. The equivalence classes, $\langle v_i \rangle = \{v_{i'} \in V' \mid v_{i'} \sim v_i\}$, include interior vertices as singletons, and boundary vertices either as singletons, $V'_{BS} = \{\langle v_i \rangle \mid v_i \in V'_B, \langle v_i \rangle = \{v_i\}\}$ (in case the point v_i is also boundary of M) or pairs $V'_{BP} = \{\langle v_i \rangle \mid v_i \in V'_B, \langle v_i \rangle = \{v_i, v_{i'}\}\}$ (v_i is not a boundary of M). Singleton boundary vertices V'_{BS} are aligned by reflections, namely, stay on some infinite line,

$$a_i^T u_i + b_i = 0, \quad \forall \langle v_i \rangle \in V'_{BS}, \tag{3}$$

where $0 \neq a_i \in \mathbb{R}^2$, $b_i \in \mathbb{R}$. Pairs of boundary vertices V'_{BP} are aligned by rotations, translations, or glide-reflections

$$u_i = r_{ii'} u_{i'} + t_{ii'}, \quad \forall \langle v_i \rangle \in V'_{BP}, \tag{4}$$

where $r_{ii'} \in O(2)$ and $t_{ii'} \in \mathbb{R}^2$. Note that (3) and (4) do not capture all the degrees of freedom for non-cone boundary vertices, $V'_B \setminus C'$; there is still one degree of freedom for singleton boundary vertex and two degrees of freedom for a pair of boundary vertices. These degrees of freedom are used to assure the convex combination property is preserved across the boundary of M':

$$\sum_{j \in \mathfrak{N}_i} \omega_{ij} (a_i^{\perp})^T (u_j - u_i) = 0, \quad \forall \langle v_i \rangle \in V'_{BS}, \tag{5}$$

where $a_i^{\perp} \perp a_i$, and

$$\sum_{j \in \mathfrak{N}_i} \omega_{ij} (u_j - u_i) + \sum_{j \in \mathfrak{N}_{i'}} \omega_{i'j} r_{ii'} (u_j - u_{i'}) = 0, \quad \forall \langle v_i \rangle \in V'_{BP}. \tag{6}$$

We denote the linear system

$$\mathcal{L} = (a) + (b) + (c).$$

The simplicial map $s : M' \to \mathbb{R}^2$ is defined by the solution of this equation (existence and uniqueness proof is deferred a bit). The simplicial map $s : M' \to \Omega$ defines a unique map $f : M \to \mathcal{N}$ by

$$f(\langle x \rangle) = [s(x)], \quad \forall x \in M'.$$

Remember that $\langle x \rangle$ represents a point in M. The map is well defined since pairs $y, z \in \langle x \rangle$ are in M'_B (i.e., the boundary of M') and are mapped, due to the boundary conditions (3), (4), to points $f(y), f(z)$ in the same orbit of G, that is, $[f(y)] = [f(z)]$.

An instrumental part of the analysis of the map s (and consequently f) is to use a tiling (i.e., branched cover) mesh M'' constructed from M' as follows. Use the transformations of the symmetry group G as defined in (3)-(4) and stitch $s(M')$ to itself along boundaries to create an infinite triangulation in the plane (with no boundary); call this triangulation M'' and the image of each vertex $v_i \in V''$ is denoted, as before, $u_i \in \mathbb{R}^2$. We will keep denoting this simplicial map s and the mesh embedding in the plane $s(M'')$. Note that $s : M'' \to \mathbb{R}^2$ satisfies the convex combination property at *all* vertices. Indeed, it is clearly so (i.e., by construction) at vertices in M'' originated from interior and boundary non-cone vertices of M'. It is also true for vertices originated from cone vertices C' of M' since at these points a point sub-group of G was applied to the neighbours of every cone point and therefore the cone points are at the centroid of their neighbours in M''. Further note that M'' is a boundaryless surface triangulation and therefore 3-connected.

PROPOSITION 2. *The linear system $\mathcal{L}$ is non-singular.*

PROOF. Assume a non-trivial solution u to the *homogenous* version of the linear system $\mathcal{L}$. Non-singularity will be proved by showing that u is necessarily trivial, $u \equiv 0$. Let $h(u) = a^T u$ be an arbitrary linear functional in $\mathbb{R}^2$.

Consider the *homogeneous* tiling $s(M'')$ of $s(M')$, i.e., using the homogeneous version of (3)-(4). That is, without a translational part. Further note that u_i for all vertices originated from cone points C' are all zero. This can be understood from the homogeneous version of (2) and the homogeneous transformations (3)-(4).

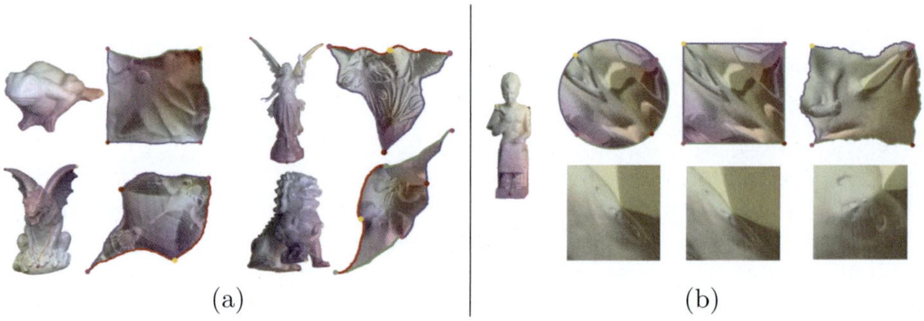

FIGURE 5. (a) shows convex combination homeomorphism of surface triangulations to different sphere-type orbifolds (from top-left): (244), (236), (333), and (2222). (b) shows convex homeomorphism to a convex polygonal domain (two left columns) and to a sphere-type (244) orbifold; bottom row shows blow-ups of the image. Note that the orbifold map is approximately conformal. *Images taken from* [**AL15**].

Lastly, note that there are only a finite number of homogeneous transformations in the symmetry groups G; in other words, the homogeneous tiling of $s(M')$ consists of infinitely repeating copies of a finite number of linear (not affine) transformations of $s(M')$.

By the comments above, the vertex function h defined by $h_i = a^T u_i$ is a convex combination function over all vertices of M''. Let $h_j = \min_i h_i$. Such a minimum exists since $s(M'')$ consists of a finite number of copies of $s(M')$. If $h_j < 0$ we can use a discrete maximum principle argument and construct a path P connecting u_j and its nearest cone vertex u_c in $s(M'')$ and conclude that $h_c = h_j < 0$, in contradiction to the homogeneous version of (2) that asserts $u_c = 0$. Repeating this argument for $h_j = \max_i h_i$ we get that $h \equiv 0$. Since a is arbitrary $u \equiv 0$. □

Figure 5(a) shows some examples of convex combination homeomorphisms of topological spheres onto a sphere-type Euclidean orbifold; (b) shows a comparison of convex combination mappings of the same cut surface M' onto two convex polygonal domains (disk-like and square) and a sphere-type orbifold. Note how angles are better preserved in the latter mapping; this will be discussed later on. Note that in case that $\mathcal{N}$ is a convex polygonal domain only equations (a) and (c) are used and there is no need for (b). That is, all boundary vertices of M are mapped to vertices of $\Omega = \mathcal{N}$.

THEOREM 1. *Let M be a 3-connected surface triangulation and $\mathcal{N} \in \mathcal{F}$ a homeomorphic target domain. Then, the simplicial map f defined by $\mathcal{L}$ is a homeomorhpism.*

3. Proof of homeomorphism

It is not a-priori clear that $f : M \to \mathcal{N}$ constructed above is a homeomorphism. Indeed, f constructed with $\mathcal{N}$ taken to be a non-convex polygonal domain or a different Euclidean cone surface in general will not always be a homeomorphism. Next we will prove the homeomorphism property of f using arguments from different papers [**AL16**, **Lov04**, **Lip14**]. We will prove only the case of $\mathcal{N}$ being a orbifold for two reasons: first, the proof for convex polygonal domains can be easily deduced from the proof below (it is almost a particular case), and second, the proof for convex polygonal domains has many excellent versions in the literature, e.g., the concise proof in [**EH10**], and the proof based on a discrete index theorem in [**GGT06**]. We also note the proof in [**Lov04**]. In the original orbifold paper [**AL15**] the proof of the embedding is done by reduction to the torus case, which is a particular instance of a Euclidean orbifold and is a sub-group of all other wall-paper groups. We provide here, hopefully, a clear, short and self-contained proof. The proof follows the following steps:

(i) The simplicial map $s : M' \to \mathbb{R}^2$ defined by the solution to $\mathcal{L}$ does not degenerate and maintains the orientation of at least one triangle of M'.
(ii) Given two triangles sharing an edge in M'. If s does not degenerate one of the triangles then it also does not degenerate the other triangle and the images of the two triangles under f will be on two different sides of the common edge.
(iii) s does not degenerate and maintains the orientations of the triangles of M' and therefore defines a homeomorphism $f : M \to \mathcal{N}$.

4. Part (i)

We will consider the mesh M'' and the tiling $s(M'')$ of $s(M')$. Consider a generic point $u \in \mathbb{R}^2$ (i.e., a point not on any edge image of M'', i.e., $u \notin s(E'')$). We will show it is contained in some non-degenerate positively oriented triangle $s(f_{ijk})$. To find such a triangle we can compute the winding number $\omega(u,t)$ of u w.r.t. the oriented boundary curve of some oriented triangle $t = s(f_{ijk})$. If $u \in t$ then $\omega(u,t) = \pm 1$ otherwise $\omega(u,t) = 0$. Another property of the winding number is that

$$\sum_i \omega(u,t_i) = \omega(u, \cup_i t_i). \tag{7}$$

Let us denote the diameter of $s(M')$ by $d > 0$. To find a triangle containing u let us consider a tiling of enough copies of M' around u using group transformations G (as defined in Equations (3)-(4) and Figure 2) so that: (i) any copy of M' not considered is of distance greater than d to u; and (ii) the boundary of the union of this tiling is a closed polygonal line of distance at-least d to u. From (i)+(ii) we get that the winding number of u w.r.t. the boundary of the tiling is 1. On the other hand by (7) the winding number equals the sum of windings of all triangles in the tiling. Therefore, there has to be at-least one triangle positively oriented and containing u.

5. Part (ii)

To prove (ii) we will start with the following lemma for the infinite triangulation M'':

LEMMA 1. *Let $s(M'')$ be a tiling generated from a convex combination map $s(M')$ defined as the solution of $\mathcal{L}$. Consider an infinite line $\ell \subset \mathbb{R}^2$ and vertices u_i, u_j which are strictly on one side of ℓ. Then, there is a path $P \subset E''$ connecting u_i, u_j that is also strictly on the same size of ℓ.*

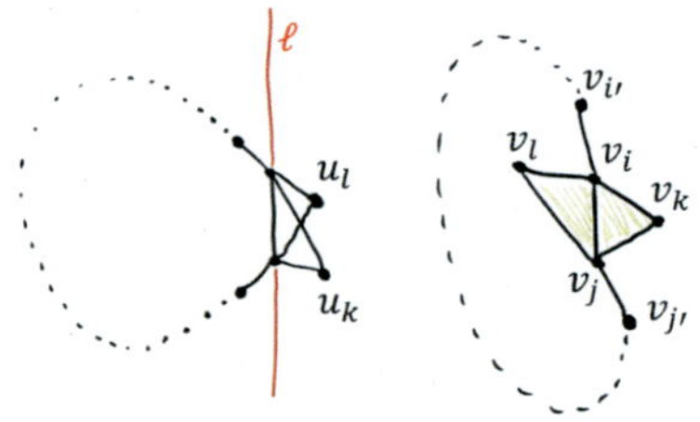

FIGURE 6. Diagram for proof of (ii) in homeomorphism proof.

Before proving this Lemma let us use it to prove (ii). Let f_{ijk}, f_{ikl} be two faces in M' sharing an edge and f_{ijk} is not degenerate. It is enough to show (ii) for a copy of these faces in M'' (since all these copies are rigid motions of the original triangles). For a bit of notational convenience we will treat $u \in \mathbb{C} \cong \mathbb{R}^2$, and $\Re(u)$ will denote the real part of u. Without loosing generality assume that $\ell = \{u \mid \Re(u) = 0\}$ and $\Re(u_k) > 0$. Assume toward contradiction that $\Re(u_l) \geq 0$, see Figure 6. By the convex combination property there are $v_{i'}, v_{j'}$ neighbors of v_i, v_j (resp.) so that $\Re(u_{i'}), \Re(u_{j'}) < 0$. Use Lemma 1 to connect $v_{i'}$ and $v_{j'}$ by a path $P = [v_{i'} = v_{p_0}, v_{p_1}, v_{p_2}, \ldots, v_{p_k}, v_{j'} = v_{p_{k+1}}]$ with $\Re(u_{p_i}) < 0$, for $i = 1, \ldots, k$. Now consider the simple closed path $Q = [v_j, v_i, P]$. The path Q divides M'' to two connected parts, one bounded and one unbounded. Consider the bounded part $U \subset V''$: it contains either v_k or v_l. Since $\Re(u_q) \leq 0$ for all $u_q \in Q$, the (second part of the) discrete maximum principle (Proposition 1) implies that $\Re(u_j) \leq 0$ for all $v_j \in U$. Since $\Re(u_k) > 0$, U cannot contain v_k.

So $v_l \in U$. In fact, v_l is in the interior of U. Indeed, $v_l \neq v_i, v_j$, and $\Re(u_l) \geq 0$ while $\Re(u_{p_i}) < 0$, $i = 0, \ldots, k+1$ so $v_l \notin P$.

There exists $U' \subset U$ a 3-connected subgraph containing v_l and a boundary point $v_b \in P$. By the 3-connectedness of M'' removing v_i, v_j keeps M'' connected and we can connect v_l to a vertex outside U using a path not passing through v_i, v_j which mean the path must intersect a vertex $v_b \in P$. Let this path be $P' = [v_l = w_0, w_1, \ldots, w_{s-1}, v_p = w_s]$. W.l.o.g. we can assume v_p is the first boundary vertex in P (since otherwise we can shorten the path). The union of the 1-neighborhoods of w_r, $r = 0, \ldots, s$ is a 3-connected graph.

Lastly, since U' is 3-connected, $\Re(u_l) \geq 0$ while there is a boundary vertex in U' so that $\Re(u_b) < 0$ we have a contradiction by the maximum principle (since $\Re(u_l) \geq 0$ attains a maximal value).

PROOF (LEMMA 1). Lets assume, as before, without loosing generality that $\ell = \{u \mid \Re(u) = 0\}$. Let us first show that we can find an infinite path $P = [v_i, v_{i_1}, v_{i_2}, \ldots]$ starting from u_i so that $\Re(u_i) \leq \Re(u_{i_k}) \nearrow \infty$. Assume that v_i has a neighbor v_{i_1} so that $\Re(u_{i_1}) > \Re(u_i)$. Then from the convex combination property we can construct a strictly monotone infinite series $P = [v_i, v_{i_1}, v_{i_2}, \ldots]$. We need to show that $\Re(u_{i_k}) \to \infty$. Since M'' is made out of finite number of types of edges there is a number $\delta > 0$ so that $\Re(u_{i_{k+1}}) - \Re(u_{i_k}) \geq \delta$ and the convergence to infinity is proven.

If v_i does not have such a neighbor, the convex combination property means all its neighbors are on the line $\ell + \Re(u_i)$. Since M'' is not contained in $\ell + \Re(u_i)$

there is some vertex $v_{i'}$ connected to v_i by a simple path so that $\Re(v_{i'}) > \Re(v_i)$. Now we can continue as above to construct P.

Let $P = [v_i, v_{i_1}, v_{i_2}, \ldots]$ and $Q = [v_j, v_{j_1}, v_{j_2}, \ldots]$ be two monotonic paths starting from v_i and v_j (resp.) and going to infinity. If P, Q intersect at some vertex we have a simple path connecting v_i, v_j. So lets assume they do not intersect. As before, let $d > 0$ denote the diameter of one copy of the image of M' in $\mathbb{R}^2$, $s(M')$. Therefore, any two cones v_{c_1}, v_{c_2}, or a vertex v_k and a cone v_{c_1} in the same copy of M', can be connected by a shortest path that is contained in a disk of diameter d. Therefore given two arbitrary vertices v_k, v_l, such that $\Re(u_k) \leq \Re(u_l)$ they can be connected by a path R going from v_k to a nearest cone v_{c_1}, traveling on the two dimensional grid made of cones to a nearest cone v_{c_2} to v_l and then to v_l. Note that the cone grids of all Euclidean orbifolds are regular and made out of two generators, $n\eta + m\xi$. All the vertices of R satisfy $\Re(u_r) > \Re(u_k) - 2d$. Therefore continuing P, Q until their distance from ℓ is at-least $2d$, they can be connected by a path R also to the right of ℓ. Concatenating P, R, Q^{-1}, possibly eliminating cycles will provide the desired path. □

6. Part (iii)

To prove (iii) we note that we have by now that all triangles in $s(M'')$ are non degenerate and positively oriented. We will repeat the winding number argument above to conclude that every generic point $u \in \mathbb{R}^2$ is covered by exactly one triangle $s(f_{ijk})$. Repeating the same argument we see the winding w.r.t. a sufficiently large tiling is 1. Also the triangles not participating in the tiling are too far to contain u. Since all the triangles in the tiling can contribute either 0 or 1, we conclude exactly one triangle contain u. Since s is an open map it means that every point u is covered exactly once. Therefore $s : M'' \to \mathbb{R}^2$ is a homeomorphism, and consequently $f : M \to \mathcal{N}$ is a homeomorphism on $\mathcal{N}$.

7. Variational principle

When the weights are symmetric, $w_{ij} = w_{ji}$, convex combination maps have variational formulation. Denote the discrete Dirichlet energy by

$$E_D(u) = \frac{1}{2} \sum_{e_{ij} \in E} w_{ij}(u_i - u_j)^2. \tag{8}$$

The linear system of equations $\mathcal{L}$ characterizes the solution of the following variational problem:

$$\begin{aligned}
&\text{(9a)} \qquad && \min_u \quad E_D(u) && \\
&\text{(9b)} && \text{s.t.} \quad a_i^T u_i + b_i = 0, && \forall \langle v_i \rangle \in V'_{BS} \\
&\text{(9c)} && \qquad u_i = r_{ii'} u_{i'} + t_{ii'}, && \forall \langle v_i \rangle \in V'_{BP} \\
&\text{(9d)} && \qquad u_i = p_i, && v_i \in C'
\end{aligned}$$

One way [**PP93, Flo03a**] to figure out a good choice of weights w_{ij} is to choose it so that (8) becomes the Dirichlet energy, $\int_M |\nabla f|^2$, when computed on a piecewise linear simplicial map f. A calculation shows that in this case

$$w_{ij} = \cot \alpha_{ij} + \cot \beta_{ij},$$

FIGURE 7. An approximation of the notorious Riemann map from a polygonal approximation of the Koch snow-flake (computed with 6 recursions) to a triangle. The approximation is computed with a 6 million vertex mesh and captures different resolutions of this map as shown in the blow-up on the right. *Image taken from* [**DSL19**].

where α_{ij}, β_{ij} are the angles opposite to the edge e_{ij} in M. These weights are called the cotan weights for obvious reasons. The cotan weights are symmetric and since $w_{ij} = \frac{\sin(\alpha_{ij}+\beta_{ij})}{\sin\alpha_{ij}\sin\beta_{ij}}$ they are positive iff the sum of angles supporting e_{ij} is smaller than π; such a triangulation M is called *Delaunay* (without 4 cocircle points). Using non-positive weights will generically damage the homeomorphism property of the map f but approximation properties will not be affected (discussed later). Another interesting option of weights which are not symmetric but always positive are the mean-value weights [**Flo03b**]. Although they don't possess a variational principle due to the non-symmetry these weights can be used in the linear system $\mathcal{L}$ to produce homeomorphic convex combination mappings.

8. Approximation of conformal mappings

The variational representation (9) leads to an interesting observation in case w_{ij} are the cotan weights [**AL15**]. The Dirichlet energy can be seen as an upper-bound to the area functional,

$$E_D(u) \geq E_A(u),$$

where E_A denotes the area functional, measuring the sum of (positive) triangle areas in $s(M')$. In case M is 3-connected and Delaunay, $w_{ij} > 0$, and Theorem 1 implies that $E_A(u) = \text{area}(\Omega) = \text{area}(\mathcal{N})$. The difference

$$E_C(u) = E_D(u) - E_A(u)$$

can be seen as the conformal distortion, and in the case u is a homeomorphism it equals the so-called least-squares conformal energy [**LPRM02**]. When using $\mathcal{N} \in \mathcal{F}$ with three cone points the number of constrained points exactly matches the degrees of freedom specified by the Riemann mapping theorem. Therefore, one

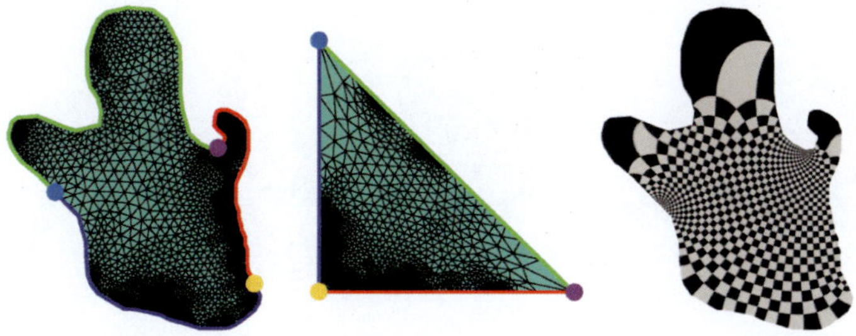

FIGURE 8. Convex combination homeomorphism, mapping a triangulation of a simply connected polygonal domain (left) onto a Euclidean orbifold triangle (middle). On the right, a pull-back checkerboard texture via the inverse of the mapping to visualize the conformality of the discrete mapping.

can ask if $f : M \to \mathcal{N}$ actually converges under refinement to the unique conformal map, mapping three prescribed points in M to the cones of $\mathcal{N}$. This was proved recently in [**DSL19**] for the case of a simply connected polygonal domain in the plane, and when $\mathcal{N} \in \mathcal{F}$ is a triangle. In fact it is shown that convergence in H^1 holds for any triangle (not just ones that can tile the plane) and any mesh (not necessarily Delaunay) however it is uniform when M is Delaunay and $\mathcal{N} \in \mathcal{F}$ (so f is a homeomorphism):

THEOREM 2. *Let $\mathcal{P} \subset \mathbb{R}^2$ be a simply-connected polygonal domain and $\mathcal{N}$ a triangle. Let M_h be surface triangulations of $\mathcal{P}$ with maximal edge length $h \to 0$ and all angles of the triangulations bounded below by some $\delta > 0$. Let $f_h : \mathcal{P} \to \mathcal{N}$ denote the simplicial maps defined by solving $\mathcal{L}$ fixing three pre-images in $\partial\mathcal{P}$ to the cones of the triangle. Let $\Phi : \mathcal{P} \to \mathcal{N}$ be the Riemann map fixing the same pre-images. Then,*

$$\|f_h - \Phi\|_{H^1} \to 0.$$

Furthermore, if M_h are Delaunay and 3-connected, then the convergence is also uniform.

Figures 7,8 show examples of approximations of Riemann mappings from polygonal domains to Euclidean orbifolds; the checkerboard texture visualizes the conformality of the simplicial mappings.

9. Surface to surface mappings

Another application of convex combination mappings is Problem 1, namely construction of a homeomorphism $f : M_1 \to M_2$ between two surface triangulations M_1, M_2. Consider a cone surface $\mathcal{N} \in \mathcal{F}$ that is topologically equivalent to both M_1 and M_2. Construct simplicial homeomorphisms $f_1 : M_1 \to \mathcal{N}$ and $f_2 : M_2 \to \mathcal{N}$ and consider $f = f_2^{-1} \circ f_1 : M_1 \to M_2$. The map f can be seen as simplicial if one considers an isomorphic joint subdivision (a.k.a. meta-mesh) M of M_1 and M_2. Using $\mathcal{N}$ with three cones provides a simplicial homeomorphism approximating the conformal map interpolating arbitrary three vertices between M_1 and M_2. (Note that M_1, M_2 are in particular Riemann surfaces.) Using $\mathcal{N} \in \mathcal{F}$ with four cones (i.e., orbifold of type (2222)) provides a quasiconformal map $f : M_1 \to M_2$ with approximately constant conformal distortion. Therefore, the orbifolds $\mathcal{N}$ with

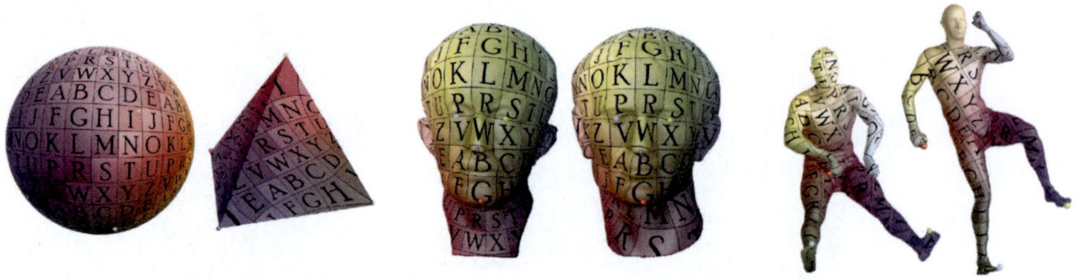

FIGURE 9. A simplicial map between pairs of surface triangulations composed out of two convex combination mappings to a common orbifold (here (2222)); the colored spheres indicate the interpolated points $(x_i, y_i) \in M_1 \times M_2$ in the map. *Images taken from* [**AL15**].

the four cone points provide a quasiconformal approximation to the Teichmüller map interpolating four points on the sphere [**AEK**$^+$**66**]. Indeed, it is possible to write $f = f_2^{-1} \circ A \circ f_1$, where f_1, f_2 are homeomophisms of finite surface triangulations, which makes them quasiconformal, and since f_1, f_2 approximate conformal mappings their conformal distortion is close to 0; A is an affine map. Figure 9 depicts examples of approximate quasiconformal mapping between pairs of surface triangulations interpolating four landmark points.

10. Other Euclidean cone surfaces

PROBLEM 2. When are convex combination mappings onto Euclidean cone surfaces not in $\mathcal{F}$ homeomorphisms?

This could be useful for discrete mappings for two reasons: (i) it will allow constraining more than three or four points; and (ii) it will potentially allow choosing $\mathcal{N}$ so to reduce isometric distortion exerted by the map. As far as the author knows there are no Euclidean cone surfaces outside the family $\mathcal{F}$ that generically allow homeomorphic convex combination maps. Furthermore, an interesting (and not completely solved) problem is:

PROBLEM 3. Find a characterization of when a Euclidean cone surface provides a homeomorphic convex combination map, even if it does not allow it generically.

Let us review several known results addressing this latter problem discussing other Euclidean cone surfaces $\mathcal{N} \notin \mathcal{F}$. In [**GGT06**] non-convex polygonal domains are discussed and a sufficient condition for a convex combination map to be bijective in this case is that every reflex cone (i.e., a cone x with angle sum $\theta(x) > \pi$) is in the convex hull of its neighbors. (Notice that this latter condition can fail.) A similar generalization to multiply-connected domains also exists.

In [**BCW17**] a Euclidean cone manifold with rational cone angles $k\frac{2\pi}{q}$, $k, q \in \mathbb{N}$ is considered and it is shown that if the triangles adjacent to the cones are positively oriented then the convex combination map is locally a homeomorphism.

In [**TFV**$^+$**13**, **AL16**] convex combination mappings are considered into the hyperbolic plane by minimizing the discrete hyperbolic Dirichlet energy [**SZS**$^+$**13**]:

$$E_D(u) = \frac{1}{2} \sum_{e_{ij} \in E} w_{ij}\, d(u_i, u_j)^2,$$

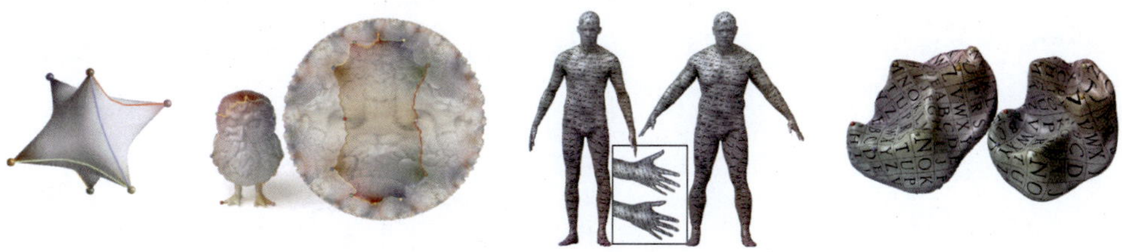

(a) (b) (c) (d)

FIGURE 10. (a) shows (an approximation) of a sphere-type hyperbolic orbifold with seven cones of angle π; (b) hyperbolic convex combination mapping of a sphere-type surface triangulation (owl) to this target orbifold; (c)-(d) homeomorphisms of surface triangulations by composing hyperbolic convex combination mappings to a common orbifold. *Images taken from* [**AL16, MGA$^+$17**].

where $u_i \in \mathcal{H}$ are points in the hyperbolic plane and $d(\cdot, \cdot)$ denotes the hyperbolic distance. Boundary conditions can be added to form a variational problem generalizing (9) to the hyperbolic case, namely for defining and computing homeomorphic simplicial mappings of a surface triangulation onto one of the hyperbolic orbifolds $\{\mathcal{N}\}$. Hyperbolic orbifolds, are defined similarly to the Euclidean orbifolds as $\mathcal{H}/G$ where G is a symmetry group of the hyperbolic plane. In contrast to Euclidean orbifolds, hyperbolic orbifolds are an infinite family of (hyperbolic) cone surfaces that include arbitrary genus surfaces and arbitrary number of cones. The boundary conditions are similar to (9b)-(9d) forcing identified boundary vertices of M' to correspond via the relevant hyperbolic isometries, namely Möbius transformations. It is proved in [**AL16**], basically following the same proof as the Euclidean case above (this time in the Klein hyperbolic model), that critical points of this non-linear problem provide homeomorphisms onto the respective hyperbolic orbifold. Figure 10(a) shows a sphere-type hyperbolic orbifold with seven cones of angle π, i.e., with symbol (2^7); (b) shows a homeomorphic hyperbolic convex combination mapping to this hyperbolic orbifold; (c) shows a homeomorphism $f = f_2^{-1} \circ f_1 : M_1 \to M_2$ between two human surface triangulations M_1, M_2 interpolating seven landmark points, where $f_i : M_i \to \mathcal{N}$, $i = 1, 2$ are hyperbolic convex combination mapping to the same hyperbolic orbifolds. Note that allowing the interpolation of seven points produces a more faithful map than the one generated with four interpolated points in the Euclidean case, Figure 9, right. Figure 10(d) shows another application of mapping anatomical surfaces (teeth).

11. Convex combination in higher dimensions

A natural question is the generalization of convex combination mapping to three dimensional simplicial complexes. Consider a tetrahedral mesh $M = (V, E, F, T)$, where $T = \{t_{ijkl}\}$ is the tetrahedra set, $t_{ijkl} = \text{hull}\{v_i, v_j, v_k, v_l\}$. Convex combination mappings can be defined as before using (1). Unfortunately, even in the most basic case of a convex polyhedron boundary conditions (assuming M is topologically a ball) the convex combination map is not guaranteed to be a homeomorphism, and counter examples were found [**DVPV03, FPT06**].

Acknowledgments

The author of this note would like to thank Noam Aigerman and Nadav Dym for good advice and proof reading.

References

[AEK+66] Lars V. Ahlfors, *Lectures on quasiconformal mappings*, Manuscript prepared with the assistance of Clifford J. Earle, Jr. Van Nostrand Mathematical Studies, No. 10, D. Van Nostrand Co., Inc., Toronto, Ont.-New York-London, 1966. MR0200442

[AL15] Noam Aigerman and Yaron Lipman, *Orbifold tutte embeddings.*, ACM Trans. Graph. **34** (2015), no. 6, 190–1. DOI 10.1145/2816795.2818099.

[AL16] ———, *Hyperbolic orbifold tutte embeddings.*, ACM Trans. Graph. **35** (2016), no. 6, 217–1. DOI 10.1145/2980179.2982412.

[BCW17] Alon Bright, Edward Chien, and Ofir Weber, *Harmonic global parametrization with rational holonomy*, ACM Transactions on Graphics (TOG) **36** (2017), no. 4, 89.

[CBGS08] John H. Conway, Heidi Burgiel, and Chaim Goodman-Strauss, *The symmetries of things*, A K Peters, Ltd., Wellesley, MA, 2008. MR2410150

[DSL19] Nadav Dym, Raz Slutsky, and Yaron Lipman, *Linear variational principle for Riemann mappings and discrete conformality*, Proc. Natl. Acad. Sci. USA **116** (2019), no. 3, 732–737, DOI 10.1073/pnas.1809731116. MR3904686

[DVPV03] Éric Colin de Verdière, Michel Pocchiola, and Gert Vegter, *Tutte's barycenter method applied to isotopies*, Comput. Geom. **26** (2003), no. 1, 81–97, DOI 10.1016/S0925-7721(02)00174-8. Thirteenth Canadian Conference on Computational Geometry—CCCG'01 (Waterloo, ON). MR1989274

[EH10] Herbert Edelsbrunner and John L. Harer, *Computational topology*, American Mathematical Society, Providence, RI, 2010. An introduction. MR2572029

[Flo97] Michael S. Floater, *Parametrization and smooth approximation of surface triangulations*, Comput. Aided Geom. Design **14** (1997), no. 3, 231–250, DOI 10.1016/S0167-8396(96)00031-3. MR1441192

[Flo03a] Michael S. Floater, *One-to-one piecewise linear mappings over triangulations*, Math. Comp. **72** (2003), no. 242, 685–696, DOI 10.1090/S0025-5718-02-01466-7. MR1954962

[Flo03b] Michael S. Floater, *Mean value coordinates*, Comput. Aided Geom. Design **20** (2003), no. 1, 19–27, DOI 10.1016/S0167-8396(03)00002-5. MR1968304

[FPT06] Michael S. Floater and Valérie Pham-Trong, *Convex combination maps over triangulations, tilings, and tetrahedral meshes*, Adv. Comput. Math. **25** (2006), no. 4, 347–356, DOI 10.1007/s10444-004-7620-5. MR2291657

[GGT06] Steven J. Gortler, Craig Gotsman, and Dylan Thurston, *Discrete one-forms on meshes and applications to 3D mesh parameterization*, Comput. Aided Geom. Design **23** (2006), no. 2, 83–112, DOI 10.1016/j.cagd.2005.05.002. MR2189438

[Lip14] Yaron Lipman, *Bijective mappings of meshes with boundary and the degree in mesh processing*, SIAM J. Imaging Sci. **7** (2014), no. 2, 1263–1283, DOI 10.1137/130939754. MR3216643

[Lov04] László Lovász, *Discrete analytic functions: an exposition*, Surveys in differential geometry. Vol. IX, Surv. Differ. Geom., vol. 9, Int. Press, Somerville, MA, 2004, pp. 241–273, DOI 10.4310/SDG.2004.v9.n1.a7. MR2195410

[LPRM02] Bruno Lévy, Sylvain Petitjean, Nicolas Ray, and Jérome Maillot, *Least squares conformal maps for automatic texture atlas generation*, ACM transactions on graphics (TOG), vol. 21, ACM, 2002, pp. 362–371.

[MGA+17] Haggai Maron, Meirav Galun, Noam Aigerman, Miri Trope, Nadav Dym, Ersin Yumer, VLADIMIR G KIM, and Yaron Lipman, *Convolutional neural networks on surfaces via seamless toric covers*, SIGGRAPH, 2017. DOI 10.1145/3072959.3073616.

[PP93] Ulrich Pinkall and Konrad Polthier, *Computing discrete minimal surfaces and their conjugates*, Experiment. Math. **2** (1993), no. 1, 15–36. MR1246481

[RG06] Jürgen Richter-Gebert, *Realization spaces of polytopes*, Lecture Notes in Mathematics, vol. 1643, Springer-Verlag, Berlin, 1996. MR1482230

[SF04] D Steiner and A Fischer, *Planar parameterization for closed 2-manifold genus-1 meshes*, Proceedings of the ninth ACM symposium on Solid modeling and applications, Eurographics Association, 2004, pp. 83–91.

[SZS+13] Rui Shi, Wei Zeng, Zhengyu Su, Hanna Damasio, Zhonglin Lu, Yalin Wang, Shing-Tung Yau, and Xianfeng Gu, *Hyperbolic harmonic mapping for constrained brain surface registration*, Proceedings of the IEEE Conference on Computer Vision and Pattern Recognition, 2013, pp. 2531–2538.

[TFV+13] Alex Tsui, Devin Fenton, Phong Vuong, Joel Hass, Patrice Koehl, Nina Amenta, David Coeurjolly, Charles DeCarli, and Owen Carmichael, *Globally optimal cortical surface matching with exact landmark correspondence*, Information processing in medical imaging: proceedings of the... conference, vol. 23, NIH Public Access, 2013, p. 487.

[Tut63] W. T. Tutte, *How to draw a graph*, Proc. London Math. Soc. (3) **13** (1963), 743–767, DOI 10.1112/plms/s3-13.1.743. MR0158387

WEIZMANN INSTITUTE OF SCIENCE, REHOVOT, 76100 ISRAEL

Proceedings of Symposia in Applied Mathematics
Volume **76**, 2020
https://doi.org/10.1090/psapm/076/00672

Conformal geometry of simplicial surfaces

Keenan Crane

1. Overview

What information about a surface is encoded by *angles*, but not *lengths*? This question encapsulates the basic viewpoint of *conformal geometry*, which studies *holomorphic* or (loosely speaking) *angle- and orientation-preserving* maps between manifolds. In the discrete setting, however, the idea of literally preserving angles leads to an interpretation of conformal geometry that is far too *rigid*: the space of discrete solutions looks far more restricted than the space of smooth solutions it hopes to model (Figure 1). Instead, one must consider how several equivalent points of view in the smooth setting lead to a number of *inequivalent* treatments of conformal geometry in the discrete setting. This activity has recently culminated in a complete *discrete uniformization theorem* for polyhedral surfaces, which beautifully mirrors the classic uniformization theorem for Riemann surfaces.

Why study discrete conformal geometry? From an analytic point of view, *smooth* conformal maps provide a strong notion of regularity, since they are complex analytic and hence have derivatives of all orders. In the discrete setting one therefore obtains one possible notion of what it means for a discrete (e.g., simpicial) map to be "regular," even though no derivatives are available in the classical sense. From a topological point of view, conformal maps help one define canonical mappings between spaces—for instance, uniformization provides an explicit map between any two conformally equivalent surfaces, by passing through a canonical domain of constant curvature. Likewise, discrete conformal geometry can be used as a starting point for constructing canonical maps between simplicial surfaces, even when they do not have compatible triangulations [**HK15**, **BCK18**]. In applications, discrete conformal geometry has become a powerful tool for digital geometry processing algorithms, since many tasks ultimately boil down to solving sparse linear systems or easy convex optimization problems. Discrete conformal maps have hence become essential for tasks ranging from regular surface remeshing [**CZ17**], to machine learning [**MGA+17**], to digital fabrication [**KCD+16**, **SC18**]. Discrete conformal geometry also arises in the context of statistical mechanics, where theories based on discrete harmonic functions [**Smi10**] and circle patterns [**Lis17**, **KLRR18**] appear to be quite natural.

2020 *Mathematics Subject Classification.* Primary 64E10; Secondary 52C26.

Key words and phrases. Discrete differential geometry, conformal geometry.

The author was supported in part by NSF Award #1717268 and a Packard Fellowship.

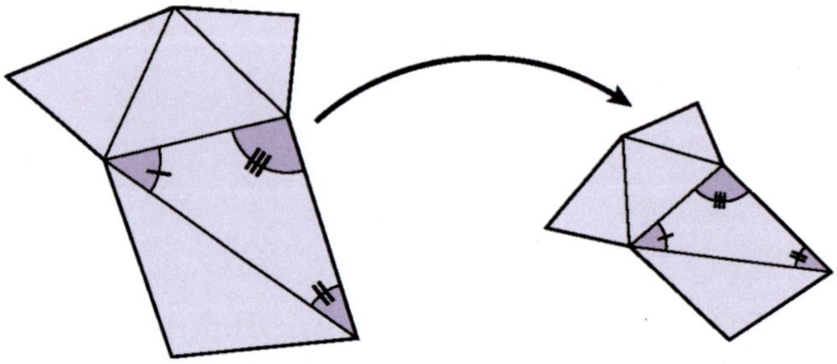

FIGURE 1. In the smooth setting, conformal maps can be characterized as maps that preserve angles and orientation. In the discrete setting, naïve applications of this characterization lead to definitions that are far too rigid: for instance, plane triangulations that share interior angles are identical up to a global similarity transformation.

Discretization of conformal maps is an excellent example of "The Game" played in discrete differential geometry [**CW17**], since there are a large number of different (yet equivalent) characterizations of conformal maps in the smooth setting, each of which leads to a distinct starting point for a discrete definition. Consider for instance a smooth, nondegenerate, and orientation-preserving map f from a disk-like surface M with Riemannian metric g to the flat complex plane $\mathbb{C}$. A reasonably comprehensive list of ways to assert that f is *conformal* includes:

(1) (**ANGLES**) Preservation of angles—at every point $p \in M$ the angle between any two tangent vectors $X, Y \in T_pM$ is the same as the angle between their images $df_p(X)$, $df_p(Y)$ in the plane.
(2) (**CIRCLES**) Preservation of circles—the image of any geodesic circle of radius ε approaches a Euclidean circle as ε goes to zero.
(3) (**ANALYTIC**) The map f satisfies the *Cauchy-Riemann equation*—$df(\mathcal{J}X) = \imath df(X)$ for all vector fields X, where $\imath$ is the complex unit and $\mathcal{J}_p : T_pM \to T_pM$ is the *linear complex structure* expressing a rotation by $\pi/2$ in each tangent space ($\mathcal{J}_p^2 = -\mathrm{id}$).
(4) (**METRIC**) Conformal equivalence of metrics—the metric g on M is related to the induced metric $\tilde{g} := df \otimes df$ via a positive scaling $\tilde{g} = e^{2u}g$, where $u : M \to \mathbb{R}_{>0}$ is called the *log conformal factor.*
(5) (**CONJUGATE**) Conjugate harmonic functions—the map f can be expressed as $f = a + b\imath$, where $a, b : M \to \mathbb{R}$ are harmonic functions with orthogonal gradients of equal magnitude ($\nabla b = \mathcal{J}\nabla a$).
(6) (**DIRICHLET**) Critical points of Dirichlet energy—among all homeomorphisms from M to $f(M)$, the map f is a critical point of the *Dirichlet energy* $E_D(f) := \int_M |df|^2 \, dA$, where dA is the area measure on (M, g).
(7) (**HODGE**) Hodge duality—the Hodge star on differential 1-forms induced by f is the same as the pushforward under f of the 1-form Hodge star on the original domain.

To someone familiar with smooth conformal geometry this list may seem redundant, since in most cases cases the conceptual leap between one characterization and another is very small. Yet these minor shifts in perspective often lead to substantially different interpretations in the discrete setting. In almost all cases, the "discrete setting" means we are considering a domain M encoded as either

(1) a triangulated surface, or
(2) a quadrilateral net.

These notes focus primarily on triangulations; see [**BG17**] and references therein for an overview of the quadrilateral case. Our aim is to give a broad overview of some of the highlights of discrete conformal geometry; many details have therefore been omitted (though see the bibliography for a detailed list of references). The notes are organized into four parts:

- **Part I** explores how a naïve treatment of discretization (e.g., preservation of angles in the triangulation, or direct discretization of the Cauchy-Riemann equation) leads to definitions of discrete conformal maps that are excessively rigid, or agree with the smooth theory only in the limit of refinement.
- **Part II** shows the first glimpse of real "discrete" theories based on preservation of circles (*circle packings* and *circle patterns*). This perspective captures many important features of conformal geometry, and very nearly provides a complete theory of discrete uniformization for general triangulations.
- **Part III** considers the perspective of conformally equivalent discrete metrics; this is to date the most satisfactory theory of conformal maps for triangulated surfaces, since it comes with a complete uniformization theorem.
- **Part IV** makes the connection between discrete conformal mapping problems and realization problems for ideal hyperbolic polyhedra, which also illuminates the relationship between theories based on circles (Part II) and conformally equivalent metrics (Part III).

Importantly, the focus in these notes on *structure preservation* should not be confused with a value judgement on *utility*: numerical schemes that do not exactly capture smooth structure may nonetheless be perfectly suitable for practical computation (especially on fine tessellations), and are often less computationally demanding than those that furnish an exact discrete theory. The same trade off can be found throughout discrete differential geometry: exact structure preservation often comes at significant computational cost relative to cheaper numerical alternatives. A negative interpretation is that one is therefore stuck between fast but inexact numerical schemes, and those that are "exact" but slow. A more mature point of view is that the two sets of tools are complementary: fast numerical methods help to initialize or approximate intermediate computations needed for exact, structure-preserving schemes, which in turn provide valuable guarantees about the behavior of algorithms. Beyond computation, discrete conformal geometry helps to bridge several areas of mathematics, including some rather remarkable connections between geometry, analysis, and combinatorics, as well as Euclidean and hyperbolic geometry. It also highlights a central thesis of discrete differential geometry, namely that the most important features of geometry are not inherently smooth nor discrete, but (as we will see) can be faithfully described in either language.

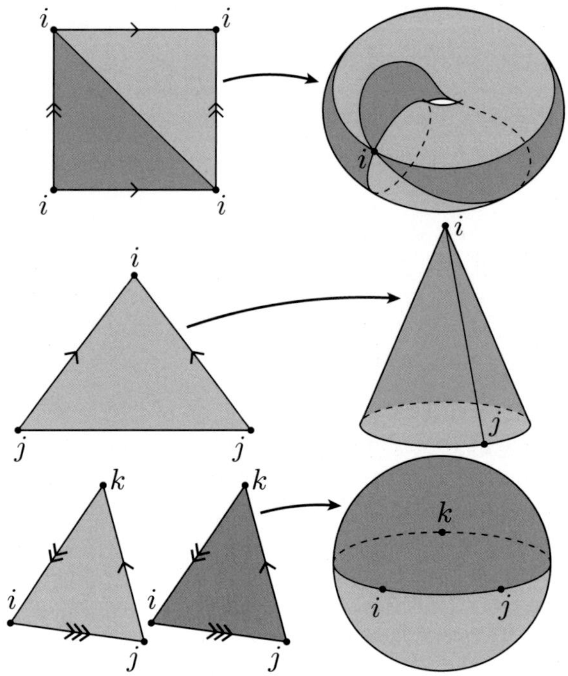

FIGURE 2. Topologically, we consider surfaces obtained by gluing triangles along edges. A torus with one vertex (top), cone with two vertices (middle), and sphere with three vertices (bottom) can each be expressed as a Δ-complex, but not as a simplicial complex.

2. Preliminaries

Throughout we consider a compact connected surface M with Riemannian metric g. In Part I we will mainly consider a differentiable map f from a disk-like domain M to the complex plane $\mathbb{C}$ (with its usual Euclidean metric); this setup allows us to focus on the local picture, and avoid issues of global topology or complex structure. As discussed in Section 1, there are many equivalent ways to express that f is *conformal*; we therefore refrain from choosing a canonical definition, and instead discuss each characterization in turn. The differential $df_p : T_p \to T_{f(p)}\mathbb{C}$ of the map f expresses how tangent vectors on M are pushed forward to $\mathbb{C}$. We will generally assume that f is an *immersion*, meaning that its differential df is nondegenerate, i.e., at each point $p \in M$, $df_p(X) = 0$ if and only if $X = 0$. In particular, every conformal map is an immersion, whereas holomorphic maps can have isolated points where the differential fails to be injective. Finally, the *linear complex structure* associated with a surface (M, g) is a tangent space automorphism $\mathcal{J} : TM \to TM$ such that in each tangent space T_pM, (i) $J_p^2 = -\text{id}$ (where id denotes the identity map), and (ii) $g_p(X, \mathcal{J}_pX) = 0$; intuitively, $\mathcal{J}$ defines a quarter-rotation in each tangent space compatible with the metric g (in analogy with the imaginary unit $\imath$ in the complex plane), and is hence determined by the metric up to a global choice of orientation.

In the discrete setting, we consider surfaces that can be obtained by gluing together Euclidean triangles—the prototypical example is a Euclidean polyhedron (such as the icosahedron), though we need not restrict ourselves to surfaces embedded in $\mathbb{R}^n$. Usually it will be sufficient to consider simplicial complexes, though for the general theory of discrete conformal equivalence (Part III) we must consider more general triangulations—here it will be sufficient to work with Δ-complexes, as defined by Hatcher [**Hat02**, Section 2.1]. Figure 2 gives several examples of triangulations that are not simplicial but are expressible as Δ-complexes. Throughout, we will use the term *discrete surface* to mean a topological 2-manifold triangulated by either a simplicial or Δ-complex $\mathsf{M} = (\mathsf{V}, \mathsf{E}, \mathsf{F})$, where V, E, and F denote the vertices (0-cells), edges (1-cells), and faces (2-cells), resp. We will use a $(k+1)$-tuple of vertex indices to specify a k-simplex; for instance, ij is an edge with vertices $i, j \in \mathsf{V}$ and ijk is a face with vertices $i, j, k \in \mathsf{V}$. (This convention is an abuse of notation in the case of a Δ-complex, but the meaning will nonetheless be clear.) If a discrete surface has finitely many triangles, we say that it is *finite* (and hence compact). The *star* $\mathrm{St}(i)$ of a vertex $i \in \mathsf{V}$ is the subcomplex of M comprised of all simplices that contain i.

The intrinsic geometry of a discrete surface $\mathsf{M} = (\mathsf{V}, \mathsf{E}, \mathsf{F})$ can be expressed via a *discrete metric*, i.e., an assignment of positive edge lengths $\ell : \mathsf{E} \to \mathbb{R}_{>0}$ that strictly satisfy the triangle inequality in each face:

$$\ell_{ij} + \ell_{jk} > \ell_{ki} \qquad \forall\, ijk \in \mathsf{F}.$$

These edge lengths naturally define a piecewise Euclidean metric on M, obtained by constructing disjoint Euclidean triangles with the prescribed lengths and gluing them along shared edges. The resulting space has a singular Riemannian metric g that is Euclidean away from vertices, and "cone-like" in a small neighborhood around each vertex $i \in \mathsf{V}$. Any such metric g is called a *(polyhedral) cone metric*; a more formal definition is given by Troyanov [**Tro91**, p. 4].

Any discrete metric determines interior angles θ_i^{jk} at each vertex i of each triangle ijk (e.g., via the law of sines). Letting

$$\Theta_i := \sum_{ijk \in \mathsf{F}} \theta_i^{jk}$$

be the total angle around vertex i, the *cone angle*

$$\Omega_i := 2\pi - \Theta_i \tag{2.1}$$

measures how close the vertex is to being Euclidean, providing a discrete analogue for Gaussian curvature. Intuitively, the intrinsic geometry looks like a circular Euclidean cone, or more generally, a circular wedge of the Euclidean plane glued together at opposite edges (Figure 3). Likewise, for any vertex i on the domain boundary, the angle $\mathsf{k}_i := \pi - \sum_{ijk \in \mathsf{F}} \theta_i^{jk}$ provides a discrete analogue for geodesic curvature. Together, Ω and k satisfy a discrete Gauss-Bonnet theorem: $\sum_i \Omega_{i \in \mathsf{M}} + \sum_{i \in \partial \mathsf{M}} \mathsf{k}_i = 2\pi\chi(\mathsf{M})$, where $\chi(\mathsf{M}) := |\mathsf{V}| - |\mathsf{E}| + |\mathsf{F}|$ is the Euler characteristic of M.

For an embedded surface, the extrinsic geometry can be expressed as a piecewise linear (or more properly, simplexwise affine) map interpolating given vertex coordinates $\mathsf{f} : \mathsf{V} \to \mathbb{R}^n$. Any such map is a *discrete immersion* if it is locally injective, or equivalently, if for each vertex $i \in \mathsf{V}$ the restriction of f to $\mathrm{St}(i)$ is embedded [**Cer96**, Lemma 2.2]. The condition that f be a (discrete) immersion

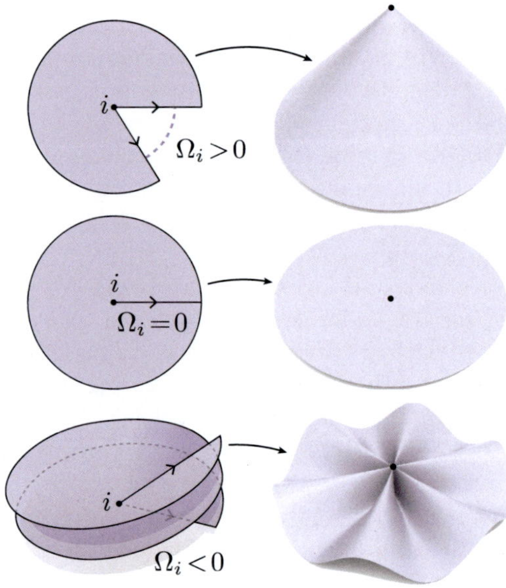

FIGURE 3. A cone metric is flat away from isolated *cone points* $i \in \mathsf{V}$, where it looks like a wedge of paper (possibly multiply-covered) identified along the two marked edges. The angle deficit or excess Ω_i determines the curvature.

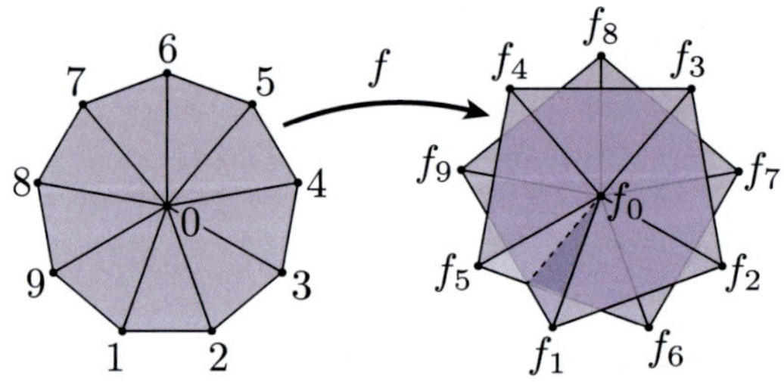

FIGURE 4. A discrete branch point, where a piecewise linear map fails to be a discrete immersion.

not only excludes vanishing angles, zero-length edges, and zero-area triangles, but also avoids discrete branch points (see Figure 4). It would therefore be reasonable to require any definition of a discrete conformal map to include the condition that f is a discrete immersion; a discrete holomorphic map should likewise be noninjective only at vertices. Any discrete immersion f induces a corresponding discrete metric

$$\ell_{ij} := |f_j - f_i|$$

(where $|\cdot|$ denotes the Euclidean norm); note that unlike the smooth setting this metric is nondegenerate even in the presence of branch points.

Since conformal maps depend only on the intrinsic geometry of a surface, the *edges* of a polyhedron have no geometric significance—even when the metric arises from an embedding in $\mathbb{R}^n$. For instance, consider an embedded polyhedron with two triangles ijk, jil sharing an edge ij (Figure 5). Since this triangle pair is isometric to a Euclidean quadrilateral (with two possible diagonals), an intrinsic

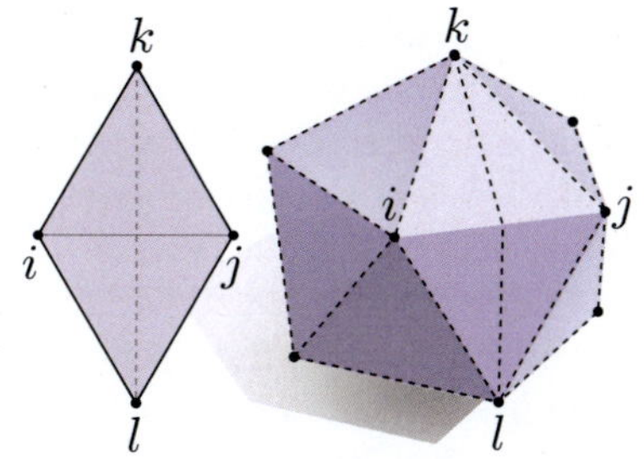

FIGURE 5. A given polyhedron admits many different geodesic triangulations, all of which describe the same intrinsic geometry.

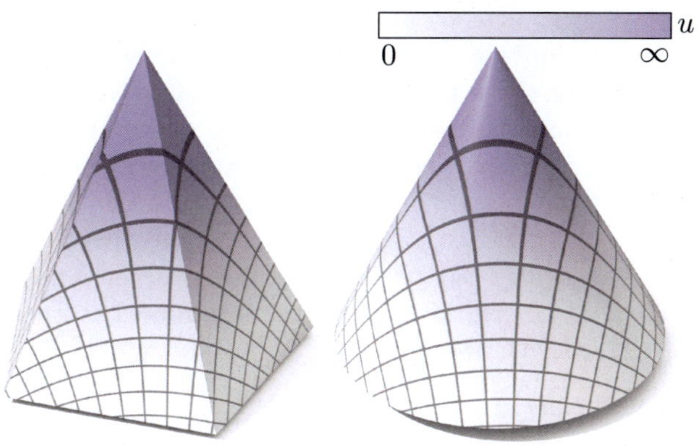

FIGURE 6. A vertex of a polyhedron (left) is intrinsically the same as a circular cone (right), and hence experiences infinite scale distortion u when conformally flattened in the classical sense. (Black lines depict pullback of integer grid lines under a conformal flattening.)

observer walking along the surface has no way to know when they are crossing the extrinsic edge. Likewise, two discrete surfaces with different triangulations and sets of edge lengths may nonetheless be isometric—consider for instance splitting the square faces of a cube along different diagonals. From this point of view, the intrinsic geometry of a discrete surface is completely determined by the number of vertices, along with their locations and cone angles. This situation motivates the construction of *intrinsic Delaunay triangulations* (Section 5.2), which play an important role in the theory of discrete uniformization (Section 5.3).

3. Part I: Discretized Conformal Maps

In Part I we consider several candidate ways to discretize conformal maps $f : (M, g) \to \mathbb{C}$ from a disk-like domain (M, g) to the complex plane $\mathbb{C}$, each of which fails to capture the behavior of smooth conformal maps in some essential way. The first question to ask, perhaps, is: why do we need a new definition at all? After all, a "discrete" surface is still a Riemannian manifold, and we should therefore be able to apply standard definitions directly (perhaps with some extra care in order to deal with cone points). This viewpoint of conformally equivalent cone metrics has in fact been studied by Troyanov [**Tro91**], though for our purposes it is not the right philosophical starting point: it views a polyhedron as a literal description of the geometry we want to represent, rather than a proxy for a smooth surface. From the classical Riemannian point of view a convex polyhedron does not look much like a smooth surface: it is flat almost everywhere, except at a discrete collection of points where it locally looks like a cone. As depicted in Figure 6, a conformal flattening of such a metric hence looks nothing like a flattening of a smooth surface, since it has unbounded scale distortion u in the vicinity of each vertex. We are therefore motivated to seek an alternative definition for discrete conformal maps, which more faithfully captures the behavior we expect from smooth surfaces.

The notions studied in this part all begin with a fairly naive hypothesis (e.g., *"discrete conformal maps should preserve angles"*), which in each case leads to a definition that is significantly more rigid than the smooth definition. An important exception is the finite element treatment discussed in Section 3.5, which provides the expected degree of flexibility, but otherwise does not capture many of the basic features of smooth conformal maps in an exact sense—for instance, even just composing a given finite element solution with a Möbius transformation may yield a map that is not the solution to any finite element problem. In all other cases, one can still consider maps which come as close as possible to satisfying the naïve definition, e.g., in the "least-squares" sense—until fairly recently, this point of view has been the starting point for most practical computational algorithms [**LPRM02**, **SLMB05**].

3.1. Interior Angle Preservation. We first consider the most basic geometric characterization on conformal maps, namely that they should preserve angles. In the smooth setting, an immersion $f : M \to \mathbb{C}$ is conformal if at every point $p \in M$ the angle between any two tangent vectors $X, Y \in T_pM$ is the same as the angle between their images $df_p(X), df_p(Y)$ in the plane. A very tempting idea, then, is to seek a discrete immersion $\mathsf{f} : \mathsf{M} \to \mathbb{C}$ that preserves the interior angles θ_i^{jk} at every triangle corner. However, it quickly becomes clear that this notion of conformal equivalence is far too rigid: interior angles are preserved if and only if each triangle $ijk \in \mathsf{F}$ experiences a similarity transformation, or (intrinsically) if the new and old edge lengths in each triangle ijk are related by a scale factor $\lambda_{ijk} > 0$, i.e., $\tilde{\ell}_{ab} = \lambda_{ijk}\ell_{ab}$ for $ab \in ijk$. At first glance this idea of locally scaling the discrete metric in each triangle feels reminiscent of the scaling relationship $\tilde{g} = e^{2u}g$ between conformally equivalent smooth metrics $g, \tilde{g}$. However, since each interior edge ij is shared by two triangles ijk, jil, the scale factors $\lambda_{ijk}, \lambda_{jil}$ of any two adjacent triangles must be identical. Globally, then, any piecewise linear map that preserves all interior angles is necessarily an isometry, up to a uniform scaling by some constant factor $c \in \mathbb{R}_{>0}$. Hence, an angle-based definition of conformal equivalence is far too rigid: each equivalence class is a one-parameter family of discrete metrics, parameterized by a single constant c. In stark contrast, smooth equivalence classes are parameterized by a scalar *function*, namely the log conformal factor $u : M \to \mathbb{R}$. (Yet as we will see in Part III a different notion of metric scaling *will* ultimately lead to the right notion of discrete conformal equivalence.)

One can nonetheless consider the map that distorts angles the least, i.e., the one that minimizes the deviation of angles from their original values in the least-squares sense [**SS00**]. To do so, suppose we parameterize piecewise linear maps $\mathsf{f} : \mathsf{M} \to \mathbb{C}$ via the interior angles $\tilde{\theta}_i^{jk} > 0$ of the image $\mathsf{f}(\mathsf{M})$. These angles describe a valid

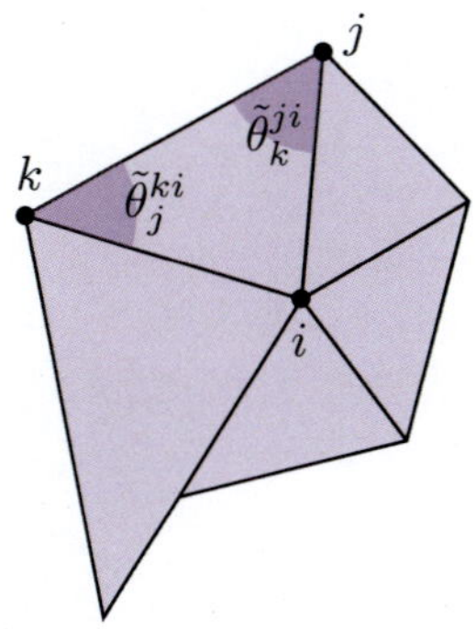

FIGURE 7. Angles that exhibit Euclidean sums around triangles and vertices may still fail to characterize a valid triangulation.

planar triangulation as long as they satisfy a collection of discrete integrability conditions:

(1) For each triangle $ijk \in \mathsf{F}$, $\tilde\theta_i^{jk} + \tilde\theta_j^{ki} + \tilde\theta_k^{ij} = \pi$.
(2) For each interior vertex $i \in \mathsf{V}$, $\sum_{ijk\in \mathrm{St}(i)} \tilde\theta_i^{jk} = 2\pi$.
(3) For each interior vertex $i \in \mathsf{V}$, $\prod_{ijk\in \mathrm{St}(i)} \frac{\sin \tilde\theta_j^{ki}}{\sin \tilde\theta_k^{ij}} = 1$.

The first condition is the usual Euclidean triangle postulate; the second is simply the requirement that every vertex have a Euclidean angle sum. As illustrated in Figure 7, the third condition is a necessary closure condition on edge lengths around each vertex, obtained by applying the law of sines to the relationship

$$\prod_{ij\in \mathrm{St}(i)} \frac{\tilde\ell_{i,j}}{\tilde\ell_{i,j+1}} = 1,$$

where $\tilde\ell_{i,j}$ and $\tilde\ell_{i,j+1}$ denote the lengths of consecutive edges in $\mathrm{St}(i)$ with respect to counter-clockwise ordering. The best approximation of an angle-preserving map (in the least-squares sense) is then any minimizer of the energy

$$E_{\mathrm{ang}}(\tilde\theta) := \sum_{ijk\in\mathsf{F}} (\tilde\theta_i^{jk} - \theta_i^{jk})^2,$$

subject to the condition that the new angles $\tilde\theta$ are positive and satisfy the discrete integrability conditions outlined above. These angles determine a discrete metric (up to a global uniform scaling), which in turn determines a discrete map to the plane (up to Euclidean motions). Here again we observe too much rigidity in the sense that E_{ang} typically has only a single unique minimizer (up to Euclidean motions), whereas in the smooth setting there is a large family of conformal maps from any disk-like domain to the plane. However, this minimizer will at least be a discrete immersion (due to condition (2) and the positive angle condition), and will exhibit variable rather than uniform scaling. Further properties of such maps are not well-understood—empirically, for instance, the minimizer of E_{ang} appears to approximate the conformal map of least *area* distortion, though this observation has never been carefully analyzed. A variety of numerical algorithms for computing such maps have been developed [**SS00, SLMB05, ZLS07**].

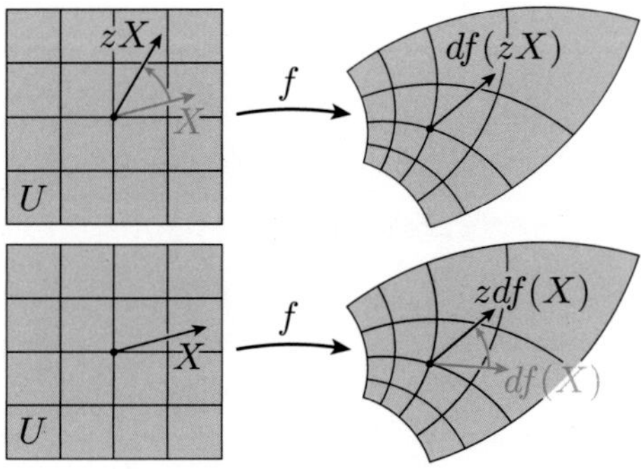

FIGURE 8. Geometrically, the Cauchy-Riemann equation asks that pushing forward vectors via given map f commutes with scaling and rotation.

3.2. Cauchy-Riemann equation. Analytically, smooth conformal maps on a domain $U \subset \mathbb{C}$ are traditionally characterized by the *Cauchy-Riemann equation*, which says that a map $f : U \to \mathbb{C}; (x, y) \mapsto u(x, y) + \imath v(x, y)$ is holomorphic if

$$\frac{\partial u}{\partial x} = \frac{\partial v}{\partial y} \quad \text{and} \quad \frac{\partial u}{\partial y} = -\frac{\partial v}{\partial x}. \tag{3.1}$$

If f is also an immersion (i.e., f has a nonvanishing derivative), then it is *conformal*. On regular quadrilateral nets, where one has clear "x" and "y" directions, this coordinate description is a natural starting point for structure-preserving discretizations of holomorphic maps that capture much of the rich structure found in complex analysis [**BG17**]. For general triangulated surfaces, where there are no clearly distinguished coordinate directions, one must take a different approach. An alternative way to write Equation (3.1) is

$$df(\imath X) = \imath df(X),$$

where $\imath$ denotes the imaginary unit and X is any tangent vector field on U. In other words, a complex map f is holomorphic if pushing forward vectors commutes with 90-degree rotation. For a surface M with linear complex structure $\mathcal{J}$, a map $f : M \to \mathbb{C}$ is likewise holomorphic if

$$df(\mathcal{J}X) = \imath df(X) \tag{3.2}$$

for all tangent vector fields X on M.

When applied to piecewise linear maps $\mathsf{f} : \mathsf{M} \to \mathbb{C}$, this characterization again leads to a definition for discrete conformal maps that is far too rigid. In particular, the only affine maps that satisfy Cauchy-Riemann on a single triangle are Euclidean motions and uniform scaling. We therefore encounter the exact same situation as we did for angle preservation in Section 3.1: in order for affine maps to agree across shared edges, they must all exhibit identical scaling. Hence, a piecewise linear map f is holomorphic in the sense of Equation (3.2) if and only if it is a global isometry, up to a global uniform dilation.

As with angle preservation, one can nonetheless seek the piecewise linear map $\mathsf{f} : \mathsf{V} \to \mathbb{C}$ that best agrees with the Cauchy-Riemann equation. Consider for instance the energy

$$E_C(f) := \int_M |*df - \imath df|^2 \, dA, \tag{3.3}$$

which measures the L^2 residual of Equation (3.2); here, $*$ denotes the Hodge star operator on differential 1-forms. To avoid the trivial solution $f = 0$, one typically incorporates a pair of point constraints $f(a) = 0, f(b) = 1$ (for $a, b \in M$, $a \neq b$). Minimizing this energy over piecewise linear maps f leads to a standard Galerkin finite element method, yielding a corresponding discrete energy $\hat{E}_C(\mathsf{f})$ (see [**CdGDS13**, Chapter 7] for a detailed derivation); this approach was one of the earliest starting points for practical conformal surface parameterization [**LPRM02**]. As with E_{ang}, the (constrained) discrete energy $\hat{E}_C$ typically has a unique minimizer. In contrast, the smooth energy E_C has an enormous space of minimizers—consider composing any minimizer f with an in-plane conformal map $\eta : f(M) \to \mathbb{C}$. Hence, the unique minimizer chosen by $\hat{E}_C$ will depend in an unstable way on superficial features of the problem (such as the choice of tessellation), hinting at some of the practical difficulties with definitions that are "too rigid". For further discussion, see [**MTAD08**, Section 3] and [**SC17**, Section 2].

3.3. Critical points of Dirichlet energy. The *Dirichlet energy* of a differentiable map $f : M_1 \to M_2$ between Riemannian manifolds (M_1, g_1) and (M_2, g_2) is the functional

$$E_D(f) := \int_{M_1} |df|^2 \, dV,$$

where dV is the volume measure on M_1; f is *harmonic* if it is a critical point of E_D. Purely topological conditions determine whether a harmonic map is also holomorphic:

THEOREM 3.1 (Eells & Wood 1975). *If $f : M_1 \to M_2$ is a harmonic map and*

$$\chi(M_1) + |\deg(f)\chi(M_2)| > 0,$$

then f is either holomorphic or antiholomorphic (where $\chi(M)$ is the Euler characteristic of M, and $\deg(f)$ *is the topological degree of f).*

Antiholomorphic simply means angle preserving and orientation *reversing*, rather than orientation *preserving*. For instance, any harmonic map from the sphere to itself automatically preserves angles, since in this case $\chi(M_1) = \chi(M_2) = \chi(S^2) = 2$. Likewise, any harmonic map f from a topological disk M to the unit disk $D^2 \subset \mathbb{C}$ will be angle-preserving, since here $\chi(M) = \chi(D^2) = 1$. Hence, we can attempt to define discrete holomorphic maps as those that minimize some kind of discrete Dirichlet energy.

For a discrete surface $\mathsf{M} = (\mathsf{V}, \mathsf{E}, \mathsf{F})$ with metric $\ell : \mathsf{E} \to \mathbb{R}_{>0}$, the discrete Dirichlet energy of a map $\mathsf{f} : \mathsf{V} \to \mathbb{C}$ can be defined as

$$\widehat{E}_D(\mathsf{f}) := \sum_{ij \in \mathsf{E}} w_{ij} |f_j - f_i|^2 \tag{3.4}$$

for any collection of edge weights w_{ij} which make this energy a positive-semidefinite quadratic form (i.e., $\widehat{E}_D(\mathsf{f}) \geq 0$ for all piecewise linear maps f). Letting k and l be the vertices opposite edge ij, a common choice is the *cotangent weights* $w_{ij} := \frac{1}{2}(\cot \theta_k^{ij} + \cot \theta_l^{ji})$, which arise from the finite element discretization mentioned in Section 3.2. (Other choices are also possible—see for instance [**HS15**].) A discrete

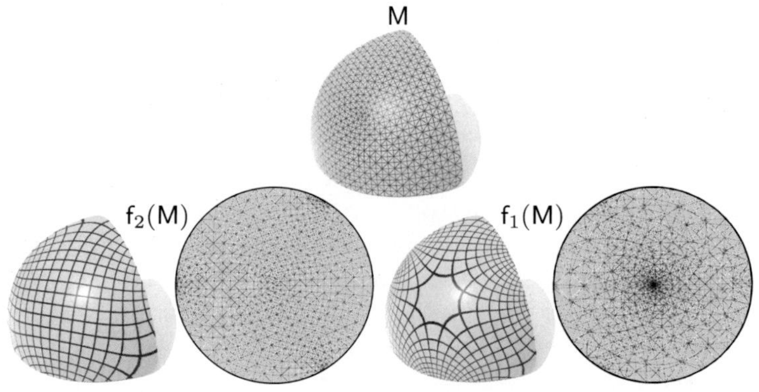

FIGURE 9. Top: two critical points $\mathsf{f}_1, \mathsf{f}_2$ of discrete Dirichlet energy among piecewise linear maps f from a triangulated disk M to the unit circular disk in the plane. Bottom: maps visualized by pulling back coordinate lines under $\mathsf{f}_1, \mathsf{f}_2$.

holomorphic map to a fixed region $U \subset \mathbb{C}$ can then be defined as a critical point of $\widehat{E}_D$, subject to the condition that boundary vertices are mapped to ∂U; this map will look conformal (rather than just holomorphic) if the boundary polygon has a unit turning number—see Figure 9 for two examples. This point of view was originally considered by Hutchinson [**Hut91**]; see also [**PP93**].

A closely related point of view is that, in the smooth setting, the conformal energy $E_C(f)$ of any map $f : M \to \mathbb{C}$ can be expressed as the difference between the Dirichlet energy $E_D(f)$ and the signed area $\mathcal{A}(f)$ of the image $f(M)$, i.e.,

$$E_C(f) = E_D(f) - \mathcal{A}(f)$$

(see [**CdGDS13**, Chapter 7]). In the case where the target is fixed (e.g., if one considers only maps to the unit disk), the area term $\mathcal{A}$ is constant, and hence E_D will have the same minimizers as E_C. One can hence define a discrete conformal map f as a minimizer of the discrete Dirichlet energy $\widehat{E}_D$ (Equation (3.4)) minus the signed area of the target polygon (subject to the same point constraints as in Section 3.2), giving another quadratic form in f:

$$\widehat{\mathcal{A}}(\mathsf{f}) := \tfrac{1}{2} \sum_{ij \in \partial \mathsf{M}} \mathsf{f}_i \times \mathsf{f}_j;$$

here $\partial\mathsf{M}$ denotes the collection of oriented edges in the boundary of M, and $z_1 \times z_2 := \mathrm{Im}(\bar{z}_1 z_2)$. This formulation is used as the starting point for several practical algorithms in digital geometry processing [**DMA02, MTAD08**]. It turns out, however, that the energy $\widehat{E}_D - \widehat{\mathcal{A}}$ is identical to the energy $\hat{E}_C$ obtained by discretizing the Cauchy-Riemann equations [**CSD02**]. Its minimizers therefore exhibit the same degree of rigidity as before.

3.4. Hodge duality. Consider the Hodge star $*$ on differential 1-forms α, which on a smooth surface with linear complex structure $\mathcal{J}$ can be expressed via the relationship[1]

$$*\alpha(X) = \alpha(\mathcal{J}X)$$

[1]Note that some authors adopt an opposite orientation convention, i.e., $*\alpha(X) = -\alpha(\mathcal{J}X)$.

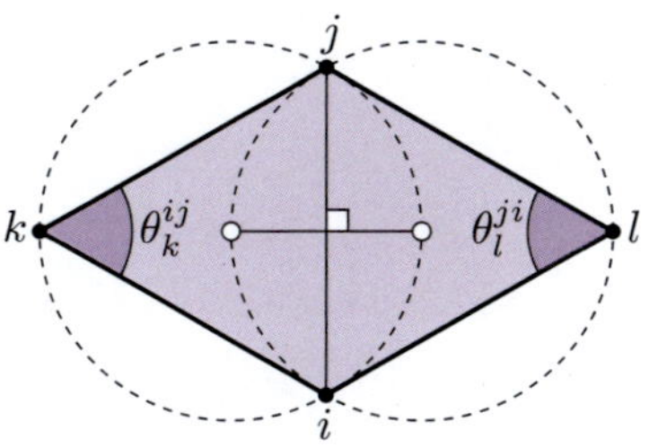

FIGURE 10. The ratio of dual over primal edge length is equal to half the sum of the cotangents of the opposite angles θ_k^{ij}, θ_l^{ji}.

for all tangent vector fields X. Since conformal maps preserve the linear complex structure, they also preserve the 1-form Hodge star. In the discrete setting, one can reverse this relationship and try to *define* a discrete conformal map as one that preserves the (discrete) Hodge star [**Mer01**]. In particular, differential forms can be discretized as cohains on a simplicial manifold and its polyhedral dual [**Wil05**, **DHLM05**]. A fairly common discretization of the Hodge star is then a "diagonal" linear map from primal k-cochains to dual $(n-k)$-cochains determined by the ratio of primal and dual volumes. In the particular case of discrete 1-forms on a triangulated surface and its circumcentric dual, a discrete differential 1-form can be encoded via a value α_{ij} per primal edge, and the corresponding (Hodge) dual 1-form can be expressed as a value on each dual edge via the cotangent formula:

$$*\alpha_{ij} := \underbrace{\tfrac{1}{2}(\cot\theta_k^{ij} + \cot\theta_l^{ji})}_{w_{ij}}\,\alpha_{ij}, \tag{3.5}$$

where θ_k^{ij}, θ_l^{ji} are the angles opposite ij, as depicted in Figure 10. A discrete conformal map in the sense of Hodge duality is then any piecewise linear map for which the new angles $\tilde\theta$ satisfy $\cot\tilde\theta_k^{ij} + \cot\tilde\theta_l^{ji} = 2w_{ij}$, or, from an intrinsic point of view, any new assignment of edge lengths $\tilde\ell_{ij}$ that preserves this same quantity.

Initially, one might be optimistic that a discretization based on the Hodge star yields less rigidity than one based on preservation of angles, since (at least naïvely) preservation of the edge weights w_{ij} effectively places only $|\mathsf{E}| \approx 3|\mathsf{V}|$ conditions on the map f, whereas preservation of interior angles θ_i^{jk} corresponds to $3|\mathsf{F}| \approx 6|\mathsf{V}|$ constraints. However, recent analysis [**ZGLG12**, **GMMD14**] extinguishes any such optimism:

THEOREM 3.2. *The primal-dual length ratios* $w_{ij} := \frac{1}{2}(\cot\alpha_{ij} + \cot\beta_{ij})$ *on a given discrete surface* $\mathsf{M} = (\mathsf{V}, \mathsf{E}, \mathsf{F})$ *uniquely determine the discrete metric* $\ell : \mathsf{E} \to \mathbb{R}$*, up to global scaling.*

The proof relies on the fact that a metric exhibiting a prescribed Hodge star can be obtained as a minimizer of a convex function, where the Hessian at any such minimizer has a kernel consisting only of vectors corresponding to global scaling of ℓ—exactly the same degree of rigidity as with angle preservation (Section 3.1), Cauchy-Riemann (Section 3.2), and Dirichlet energy (Section 3.3).

3.5. Conjugate harmonic functions. Suppose we express a holomorphic map $f : M \to \mathbb{C}$ as $f = a + b\imath$ for a pair of real-valued functions $a, b : M \to \mathbb{R}$. A straightforward consequence of the Cauchy-Riemann equation (Section 3.2) is that

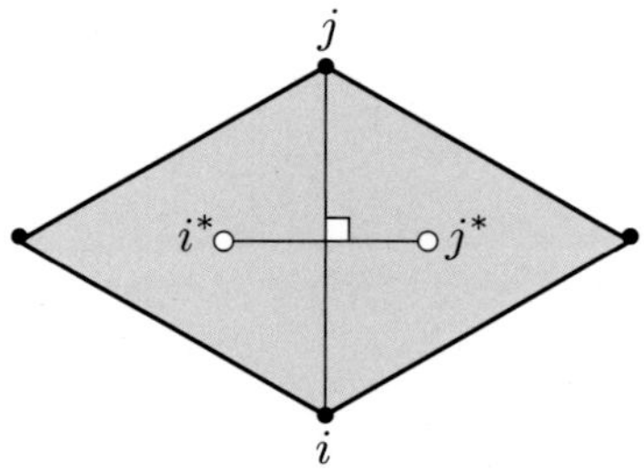

FIGURE 11. One possible notion of discrete conjugacy: a function on primal (black) vertices is conjugate to a function on dual (white) vertices if the difference across primal edges is equal to the difference across dual edges, up to a constant factor that accounts for triangle shape.

a and b form a *conjugate harmonic pair*, i.e., they are real harmonic functions with orthogonal gradients:

$$\begin{aligned} \Delta a &= 0, \\ \Delta b &= 0, \\ \nabla b &= \mathcal{J}\nabla a, \end{aligned} \tag{3.6}$$

Here Δ is the Laplace-Beltrami operator on M, ∇ is the gradient operator, and $\mathcal{J}$ is the linear complex structure.

The conditions in Equation (3.6) are of course closely related to (in fact, equivalent to) the condition that the map itself be a critical point of Dirichlet energy (Section 3.3), but the perspective of conjugate harmonic pairs will provide a different starting point for discretization, where boundary conditions play a key role. Characterizing conformal maps in terms of harmonic functions is also attractive from the perspective of discretization, since discrete harmonic functions are well-studied. In the simplicial setting, a discrete harmonic function $\phi : V \to \mathbb{R}$ is naturally defined as a piecewise linear function in the kernel of a discrete Laplace-Beltrami operator $\mathsf{L} : \mathsf{V} \to \mathsf{V}$ such as the *cotan Laplacian*

$$(\mathsf{L}\phi)_i := \tfrac{1}{2} \sum_{ij \in E} (\cot \theta_k^{ij} + \cot \theta_l^{ji})(\phi_j - \phi_i). \tag{3.7}$$

(See [**WMKG07**] for a more thorough discussion.) To define a conjugate harmonic pair, one then just needs a discrete notion of conjugacy.

What does it mean for two discrete harmonic functions to be conjugate? One idea, studied by Mercat and others [**Mer01**], is to consider functions on the combinatorial or *Poincaré dual* of a discrete surface $\mathsf{M} = (\mathsf{V}, \mathsf{E}, \mathsf{F})$, which associates each vertex with a 2-cell, each edge with a 1-cell, and each triangle with a 0-cell. In this setting, two real-valued functions a, b on the primal and dual 0-cells (resp.) are discretely conjugate if for each edge e and corresponding dual edge e^*, the difference of a values across e is equal to the difference of b values across e^*, up to a scale factor $w_e \in \mathbb{R}$ that accounts for the geometry of the triangulation (as discussed in Section 3.4). In other words, if

$$\mathsf{b}_{j^*} - \mathsf{b}_{i^*} = w_{ij}(\mathsf{a}_j - \mathsf{a}_i)$$

for all edges $ij \in \mathsf{E}$, where i^*, j^* are the associated dual vertices (see Figure 11). In terms of (discrete) differential forms, we are simply requiring that $db = *da$, capturing the conjugacy condition $\nabla b = \mathcal{J}\nabla a$. Here one encounters two basic sources of difficulty. First, as detailed in Section 3.4, if one views the conformal structure of a discrete surface as being determined by the edge weights w_{ij} (or *discrete Hodge star*), then one encounters severe rigidity: the weights uniquely determine a discrete metric. Moreover, pairs of functions a, b that are conjugate in this sense do not define a *simplicial* mapping from M to $\mathbb{C}$, since the coordinates b are associated with 0-cells of the dual complex, rather than vertices of the original discrete surface M.

Alternatively, one can take the following approach: given a discrete harmonic function a (i.e., a piecewise linear function satisfying $\mathsf{La} = 0$), its harmonic conjugate can be defined as the function $\mathsf{b} : \mathsf{V} \to \mathbb{R}$ that minimizes the L^2 difference between $\mathcal{J}\nabla\mathsf{a}$ and $\nabla\mathsf{b}$ over the space of piecewise linear functions—or equivalently, which minimizes the discrete conformal energy $\widehat{E}_C(\mathsf{f}) = \widehat{E}_D(\mathsf{f}) - \widehat{A}(\mathsf{f})$ of a map $\mathsf{f} = \mathsf{a} + \mathsf{b}\imath$ (Section 3.3), while keeping a fixed. The minimizer is the solution to the discrete Laplace equation $\mathsf{Lb} = 0$ with discrete Neumann boundary data

$$\mathsf{h}_i = \frac{1}{2}(a_{i+1} - a_{i-1}),$$

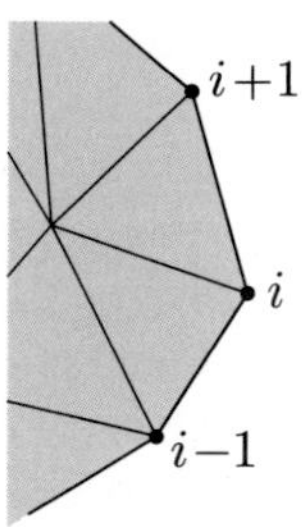

where $i-1$, i, and $i+1$ denote consecutive vertices along the boundary (see [**SC17**, Section 4.3.3] for more detail). The resulting map $\mathsf{f} := \mathsf{a} + \mathsf{b}\imath$ is then discrete conformal in the same finite element sense discussed in Section 3.2. Importantly, however, one finally obtains exactly the right amount of flexibility: whereas a "least squares" conformal map [**LPRM02**] of the kind discussed in Section 3.2 is determined up to Euclidean motions, one now obtains a whole family of discrete conformal maps f parameterized by the harmonic function a. Since this function is in turn determined by its own boundary values, one ends up in precisely the same situation as in the smooth setting, where holomorphic maps $f : M \to \mathbb{C}$ can be parameterized by real functions on ∂M specifying either Dirichlet or Neumann boundary conditions. A more geometric view is that one can parameterize such maps by specifying either the log conformal factor u or the geodesic curvature κ along the boundary—Sawhney & Crane [**SC17**] outline a complete strategy in the piecewise linear setting.

Ultimately, it should come as no surprise that a proper finite element treatment should faithfully capture the behavior of smooth conformal maps *in the limit of refinement.* On the other hand, for any *fixed* triangulation, such maps do not exactly preserve many basic properties from the smooth setting, such as covariance with respect to Möbius transformations. Moreover, they do not provide a natural notion of equivalence—for instance, the composition of two piecewise linear harmonic functions is not in general harmonic. We will therefore continue in Parts II and III to seek a precise notion of discrete conformal equivalence for finite triangulations.

4. Part II: Circle preservation

A linear map preserves angles if and only if it is the composition of a rotation and a dilation (no shear); hence, it also preserves circles. Since at each point $p \in M$ the differential df_p of a conformal map f is an angle-preserving linear map,

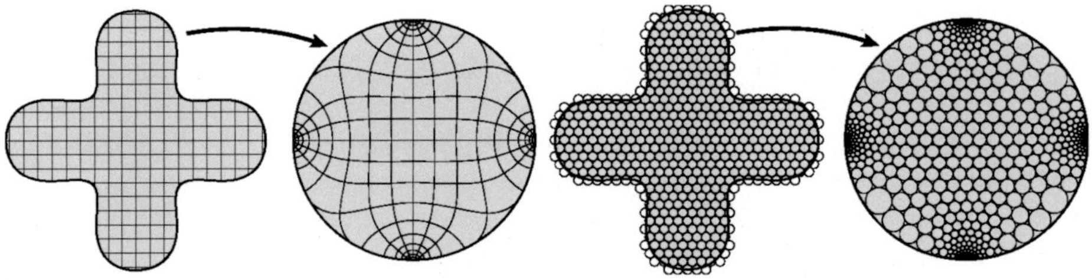

FIGURE 12. Left: a smooth Riemann mapping from a simply connected region in the plane to the unit circular disk. Right: a discrete Riemann mapping, expressed as a *circle packing*.

it must also preserve infinitesimal circles. This point of view is the starting point for several distinct but closely-related approaches to discrete conformal maps, based on arrangements of circles with special combinatorial and geometric relationships.

Finite arrangements of circles provide fertile soil for discrete conformal maps, capturing many features of the smooth theory. Koebe originally showed how incidence relationships in planar graphs can be captured via tangency relationships between circles, providing a basic connection between combinatorics and geometry. Thurston conjectured (and Rodin & Sullivan later proved) that these *circle packings* have a deep connection to Riemann mappings in the complex plane (Section 4.1); such connections have since been studied intensively by Schramm [**Roh11**] and many others. To generalize this construction to curved surfaces and irregular triangulations one must incorporate additional geometric information, such as intersection angles or other data describing relationships between circles; such arrangements fall under the general heading of *circle patterns*[2] (Section 4.2).

There are however some missing features, most notably the lack of a uniformization theorem that guarantees the existence of a constant curvature circle pattern equivalent to any given triangulation. In Part III we will see a discrete uniformization theorem which guarantees existence by considering variable rather than fixed triangulations; whether one can take an analogous approach in the circle pattern setting remains to be seen. Note that circle packings and patterns have been studied extensively beyond the context of triangulations and discrete conformal maps; see for instance [**Ste05, BS04a, BHS06, WP11**].

4.1. Circle packings. A *circle packing* is a collection of closed circular disks in the plane (or other surface) that intersect only at points of tangency—see [**Ste03**] for an excellent overview. To any such collection one can associate a graph $G = (\mathsf{V}, \mathsf{E})$ (called the *nerve*) where each vertex corresponds to a disk, and two vertices are connected by an edge if and only if their associated disks are tangent. A natural question to ask is: which graphs admit circle packings?

THEOREM 4.1 (Circle Packing). *Every planar graph* $G = (\mathsf{V}, \mathsf{E})$ *can be realized as a circle packing in the plane.*

[2]There is some inconsistency in the use of *packings* versus *patterns* throughout the literature—for clarity we adopt the convention that *packings* are arrangements of circles meeting tangentially, whereas *patterns* describe any arrangement of circles (possibly overlapping or disjoint), whether associated with vertices, faces, etc.

FIGURE 13. Discrete surfaces with identical combinatorics but different edge lengths (left) yield the same class of circle packings into the unit disk (right).

(For nonplanar graphs, one can also consider circle packings on surfaces of higher genus.) A first hint that circle packings are connected to discrete conformal maps comes via the following theorem:

THEOREM 4.2 (Koebe). *If G is a connected maximal planar graph, then it has a unique circle packing up to Möbius transformations and reflections.*

(A constructive algorithm for computing such packings was given by Collins & Stephenson [**CS03**].) A graph is *maximal* if the addition of any edge makes it nonplanar. If G is also finite, one quickly sees that it is equivalent to a triangulation of the sphere, via stereographic projection. Koebe's theorem therefore says that any triangulated sphere can be realized as a family of planar circle packings parameterized by Möbius transformations, just as smooth conformal maps from the sphere to the plane have Möbius symmetry. An even richer connection between circle packings and conformal maps can be made via the *Riemann mapping theorem*:

THEOREM 4.3 (Riemann mapping). *Any nonempty simply connected open set $\Omega \subsetneq \mathbb{C}$ can be mapped to the interior of the unit circular disk $D^2 := \{z \in \mathbb{C} : |z| < 1\}$ by a bijective map $\phi : \Omega \to D^2$ that is holomorphic (hence conformal) in both directions.*

The circle packing analogue of Riemann mapping was originally conjectured by Thurston, and later proved by Rodin and Sullivan [**RS87**]. The basic idea is to start with a regular hexagonal circle packing C of a simply connected region Ω by disks of radius $\varepsilon > 0$, i.e., for any hexagonal tiling of the plane, take only those disks that intersect Ω. Now find a circle packing C' that maintains the same incidence relationships, but where all disks along the boundary are now tangent to the unit circular disk D^2 (this idea is illustrated in Figure 12, right). The relationship between these two packings defines an *approximate mapping* of Ω to D^2: for sufficiently small ε any point $z \in \Omega$ will be contained in a circle c from C, and can be mapped to the center of the corresponding circle c' from C'. Rodin and Sullivan show that the approximate mapping converges to a conformal homeomorphism as $\varepsilon \longrightarrow 0$.

Unlike smooth Riemann mappings, however, one cannot directly use circle packings to define discrete conformal maps between any two disk-like regions, since finite hexagonal packings of these regions will not in general have the same combinatorics. More importantly, this theory does not provide a general approach to uniformization, since it cannot account for the curvature of the domain, nor triangulations with nonuniform edge lengths. Consider for instance simplicial disks with identical combinatorics but different discrete metrics $\ell : \mathsf{E} \to \mathbb{R}_{>0}$, as depicted in Figure 13. Since a circle packing depends purely on the combinatorics of the edge graph, these surfaces are realized by identical families of circle packings—implying that all discrete metrics on a given simplicial disk are, effectively, conformally equivalent. In this sense, the basic theory of circle packings is *too flexible*; one is therefore prompted to enrich it with additional data.

4.2. Circle patterns. A more direct way to encode the geometry of the domain is to incorporate additional metric information into the arrangement of circles. Suppose we associate each edge $ij \in \mathsf{E}$ of a triangulation $\mathsf{M} = (\mathsf{V}, \mathsf{E}, \mathsf{F})$ with an angle $\omega_{ij} \in [0, \frac{\pi}{2}]$; the pair (M, ω) is called a *weighted triangulation.* If we now assign a radius $r_i > 0$ to each vertex $i \in \mathsf{V}$, then a pair of circles in the Euclidean plane with the prescribed intersection angle ω_{ij} and radii r_i, r_j will have centers separated by a distance

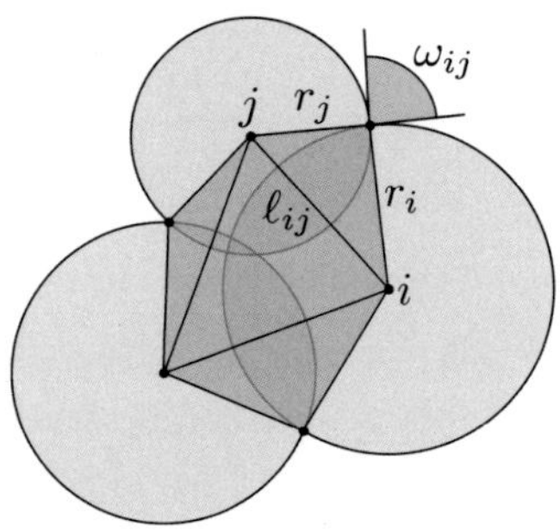

$$\ell_{ij} = \sqrt{r_i^2 + r_j^2 + 2r_i r_j \cos\omega_{ij}}. \tag{4.1}$$

The condition $\omega \leq \pi/2$ ensures that these lengths satisfy the triangle inequality in each face $ijk \in F$ (see [**Thu79**, Lemma 13.7.2]); they hence determine a discrete metric on the surface (called the *circle packing metric*), and in turn, a corresponding cone angle Ω_i at each vertex (Equation (2.1)). We can make an analogy between each of these quantities and data from the smooth setting:

- the edge lengths ℓ_{ij} encode the metric,
- the cone angles Ω_i encode the Gaussian curvature,
- the intersection angles ω_{ij} capture the conformal structure, and
- the radii r_i (or more precisely, any subsequent *changes* to the radii) play the role of conformal scale factors.

The last two items on this list can be understood by considering that (i) angles are conformal invariants, and (ii) adjusting the radii (scale factors) will scale the edge lengths ℓ (metric) while keeping the angles ω (conformal structure) fixed. From this point of view, finding a conformal map from a curved surface ($\Omega \neq 0$) to a flat one ($\Omega_i = 0$ at all vertices) amounts to finding an appropriate adjustment of radii, i.e., finding the conformal scale factors that flatten the metric. Note however that it is not immediately clear that a target angle defect Ω can always be achieved—and in general it cannot. In particular, we have the following theorems [**Thu79**]:

THEOREM 4.4 (rigidity of circle patterns). *For a closed weighted triangulation* (M, ω), *an assignment of cone angles* Ω_i *to vertices uniquely determines a unique discrete metric* ℓ_{ij} if one exists—*in other words, if there are radii* r_i *at vertices* *such that (i) circles intersect at the prescribed angles ω and (ii) the edge lengths determined by Equation* (4.1) *induce the prescribed cone angles Ω, then this circle pattern is unique up to a uniform scaling of all circle radii / edge lengths.*

THEOREM 4.5 (existence of circle patterns). *Let $\mathsf{M} = (\mathsf{V}, \mathsf{E}, \mathsf{F})$ be a closed weighted triangulation with weights ω, and for any subset of vertices $\mathsf{I} \subset V$ let* $\mathrm{Lk}(\mathsf{I})$ *and F_I denote the simplicial link of I and the set of faces with vertices in I, resp. Then prescribed cone angles Ω_i can be achieved if and only if*

$$\sum_{i\in\mathsf{I}} \Omega_i + \sum_{ij\in\mathrm{Lk}(\mathsf{I})} (\pi - \omega_{ij}) > 2\pi\chi(\mathsf{F}_\mathsf{I}),$$

where χ denotes the Euler characteristic.

In short: one cannot prescribe curvature arbitrarily, but if the prescribed curvature can be achieved, then it uniquely determines the metric (up to global scaling). The condition in Theorem 4.5 is akin to Gauss-Bonnet, but much stronger: just as any metric can be uniformized in the smooth setting, one would like the existence of discrete uniformization to depend only on a basic Gauss-Bonnet condition on the curvature data Ω, and not on the data ω describing the domain itself.

Generalizations. Generalizations of the circle patterns described above provide greater flexibility. One such generalization, first studied by Bowers and Stephenson [**BS04b**], is an *inversive distance circle packing*—here we exchange the angles ω_{ij} for values $I_{ij} \in [-1, \infty)$, and define the edge lengths via the *inversive distance*

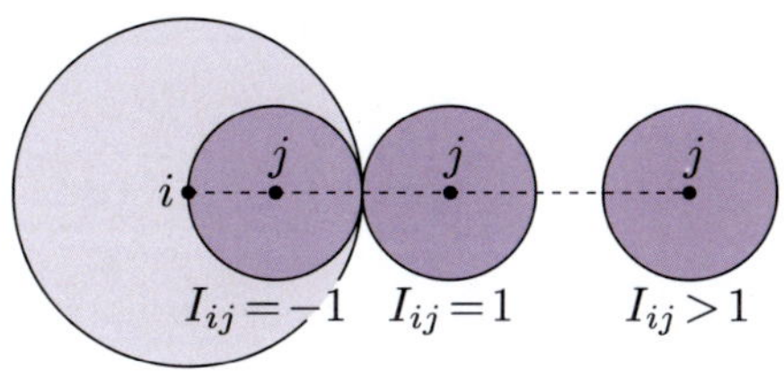

$$\ell_{ij} = \sqrt{r_i^2 + r_j^2 + 2r_ir_jI_{ij}}.$$

When $I_{ij} = 1$ circles are tangent, when $I_{ij} < 1$ they intersect at an angle $\omega_{ij} = \arccos(I_{ij})$, and when $I_{ij} > 1$ they are disjoint. In this setting one again has local and global rigidity theorems akin to Theorem 4.4 [**Guo11, Luo11, Xu18**], though no guarantee of existence due to conditions like the one from Theorem 4.5.

One can also associate a circle pattern with the *faces* of a triangulation, rather than its vertices. Let (M, ω) be a weighted triangulation where (i) the weights ω_{ij} sum to 2π around each interior vertex, and (ii) the values $\kappa_i := 2\pi - \sum_{ij} \omega_{ij}$ sum to 2π over all boundary vertices. One can then find a Euclidean circle pattern with one circle per face, so long as there is a collection of interior angles α_i^{jk} that satisfy Rivin's *coherent angle system* [**Riv94**], namely

(1) (Positivity) $\alpha_i^{jk} > 0$ at each triangle corner.
(2) (Triangle Postulate) $\alpha_i^{jk} + \alpha_j^{ki} + \alpha_k^{ij} = \pi$ for each face ijk.
(3) (Compatibility) $\pi - \omega_{ij}$ is equal to $\alpha_k^{ij} + \alpha_l^{ji}$ for each interior edge ij, and equal to α_k^{ij} for each boundary edge ij.

For instance, if ω_{ij} are the intersection angles of the triangle circumcircles in a planar Delaunay triangulation, then the circle pattern will reproduce this triangulation, up to a global similarity. Existence and uniqueness (up to similarity) can be obtained through variational arguments [**BS04a**]. However, this construction does

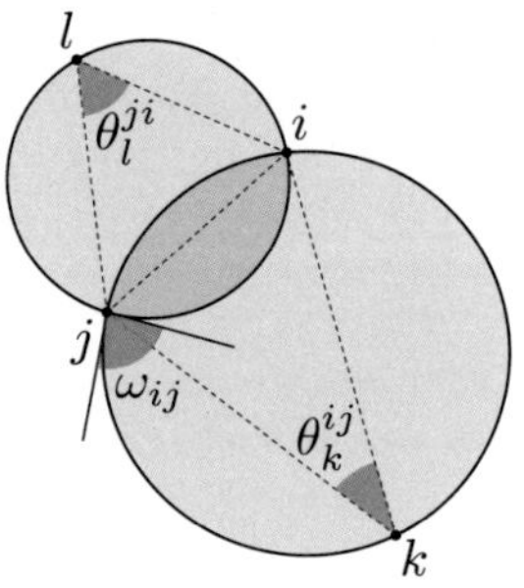

not directly address the task of uniformization, since for a curved domain the intersection angles ω_{ij} will not in general sum to 2π around interior vertices. A practical approach to conformal flattening via facewise circle patterns was proposed by Kharevych et al. [**KSS06**], who first construct a collection of interior angles α that (i) satisfy Rivin's conditions, and (ii) approximate the interior angles θ_i^{jk} of the curved domain in the least-squares sense. This particular treatment does not however provide a clean notion of discrete conformal equivalence for general (i.e., non-flat) discrete surfaces.

Other generalizations, such as *hyper-ideal circle patterns* [**Sch08, Spr08**], finally provide the right degree of flexibility. In this setting, one associates circles with both vertices *and* edges—and obtains a discrete uniformization theorem in the case of surfaces with non-positive curvature [**BDS17**], even with fixed combinatorics. Such results suggest that it is not unreasonable to expect a complete discrete uniformization theorem based on patterns of circles (namely, one that includes the positively curved case), though at present it would seem that perhaps the ending has not yet been written.

4.3. Discrete Ricci flow. In the smooth setting, one approach to uniformization is to consider Hamilton's *Ricci flow* [**Ham82**], which for surfaces can be expressed as

$$\frac{d}{dt}g = -Kg.$$

Here K denotes the Gaussian curvature of the current metric g. Intuitively, Ricci flow shrinks the metric where curvature is "too positive," and expands it where curvature is "too negative," ultimately smoothing out all bumps in curvature. Since the flow applies a pointwise rescaling at each moment in time, the final, uniformized metric is conformally equivalent to the initial one. Often one considers a normalized version of this flow

$$\frac{d}{dt}g = (\overline{K} - K)g$$

where $\overline{K}$ is the average Gaussian curvature over the domain; this flow becomes stationary when the metric has constant curvature.

Chow and Luo [**CL03**] consider a discrete analogue of this flow based on weighted triangulations (with intersection angles $\omega_{ij} \in [0, \frac{\pi}{2}]$)—in particular, they define the *(normalized) combinatorial Ricci flow*

$$\frac{d}{dt}r_i = (\overline{\Omega} - \Omega_i)r_i,$$

where the values Ω_i are the cone angles of the circle packing metric determined by r_i, and $\overline{\Omega}$ is the average cone angle. If one considers the effect of the radii r on the discrete metric ℓ, this flow has the same basic behavior as Ricci flow: the metric is rescaled in order to smooth out bumps in curvature. Chow and Luo show that this flow is defined for all time and converges to a constant curvature metric *if one exists*, i.e., the flow does not uniformize all initial metrics—the condition on existence is the same as the one given in Theorem 4.5. Nonetheless, the flow behaves well enough in practice to provide a starting point for a wide variety of algorithms in geometric computing [**GY08**]. Circle patterns can also be connected with other geometric flows such as Calabi flow [**Ge18, ZLH$^+$18**], though again Thurston's condition (Theorem 4.5) must of course hold in order to guarantee existence of a uniformized circle packing metric.

Luo also considered a closely related *combinatorial Yamabe flow* [**Luo04**], defined in terms of a different discrete analogue of conformal maps (i.e., one that is not based on circle packings or patterns). This flow provides a starting point for the notion of discrete conformal equivalence that we will study in Part III, and ultimately, to a complete discrete uniformization theorem where existence can be guaranteed.

5. Part III: Metric scaling

In the smooth setting, two Riemannian metrics $g, \tilde{g}$ on a surface M are *conformally equivalent* if they are related by a positive scaling, i.e., if $\tilde{g} = e^{2u}g$ for some function $u : M \to \mathbb{R}$ called the *log conformal factor*. Hence, e^u gives the length scaling, and e^{2u} is the area scaling. A *conformal structure* on M is then an equivalence class of metrics.

This section examines a definition for conformal equivalence of discrete metrics that resembles the one found in the smooth setting, and which ultimately allows one to formulate a discrete uniformization theorem for polyhedral surfaces (Section 5.3) that mirrors the one for smooth Riemann surfaces. From an applied point of view discrete uniformization provides a principled starting point for tasks like regular surface remeshing [**CZ17**] and constructing maps between polyhedral surfaces [**HK15**]. We begin with a basic picture involving only Euclidean geometry (Section 5.1), and will revisit this story through the lens of hyperbolic geometry in Part IV.

5.1. Conformally equivalent discrete metrics. Recall that a *discrete metric* is an assignment of edge lengths $\ell : \mathsf{E} \to \mathbb{R}_{>0}$ that satisfy the triangle inequality in each face, i.e., $\ell_{ij} + \ell_{jk} \geq \ell_{ki}$ for all $ijk \in \mathsf{F}$. What does it mean for two discrete metrics $\ell, \tilde{\ell}$ to be conformally equivalent? One idea is to simply mimic the smooth definition and ask that $\tilde{\ell}_{ij} = \lambda_{ij}\ell_{ij}$ for some scale factor λ_{ij} on each edge $ij \in \mathsf{E}$. However, this constraint is clearly "too flexible," since then every pair of discrete metrics is conformally equivalent—simply let λ_{ij} be the ratio $\tilde{\ell}_{ij}/\ell_{ij}$. Also recall from Section 3.1 that a scale factor *per face* is "too rigid," since one is then forced to make all scale factors equal. An alternative is to consider scale factors at vertices [**RW84, Luo04**]:

DEFINITION 5.1. Two discrete metrics $\ell, \tilde{\ell}$ on the same triangulation $\mathsf{M} = (\mathsf{V}, \mathsf{E}, \mathsf{F})$ are *discretely conformally equivalent* if for each edge $ij \in \mathsf{E}$,

$$\tilde{\ell}_{ij} = e^{(u_i+u_j)/2}\ell_{ij}, \tag{5.1}$$

for some collection of values $u : \mathsf{V} \to \mathbb{R}$.

This definition again gives the impression of merely aping the smooth relationship—yet in this case the resulting theory is neither too rigid nor too flexible. Instead, it provides a rich notion of discrete conformal equivalence that beautifully preserves much of the structure found in the smooth setting [**BPS15, GLSW18**], a perspective which has had a significant impact on (and has been inspired by) algorithms from digital geometry processing [**SSP08, BCGB08, CZ17**].

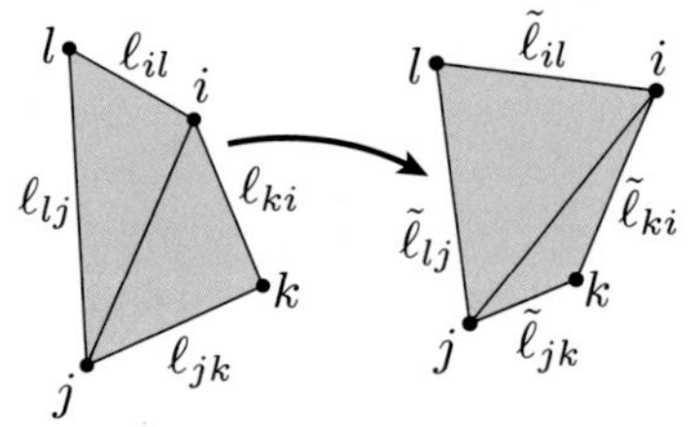

FIGURE 14. A piecewise linear map is discrete conformal if and only if it preserves the length cross ratio $\mathfrak{c}_{ij} := \ell_{il}\ell_{jk}/\ell_{ki}\ell_{lj}$ for each interior edge ij.

One elementary observation is that, *locally*, conformally equivalent discrete metrics in the sense of Equation (5.1) are still very flexible:

LEMMA 5.2. *Any two discrete metrics $\ell, \tilde{\ell}$ on a single triangle ijk are discretely conformally equivalent.*

PROOF. The metrics are conformally equivalent if there exists an assignment of log scale factors u_i, u_j, u_k to the three vertices such that

$$e^{(u_a+u_b)/2} = \tilde{\ell}_{ab}/\ell_{ab}$$

for each edge $ab \in ijk$. Let $\lambda_{ij} := 2\log(\ell_{ij})$ (and similarly for $\tilde{\ell}$). Then by taking the logarithm of the system above we obtain a linear system for the u values, namely

$$u_a + u_b = \tilde{\lambda}_{ab} - \lambda_{ab}, \quad \forall ab \in ijk.$$

This system has a unique solution, independent of the values of ℓ and $\tilde{\ell}$. In particular,

$$e^{u_i} = \frac{\tilde{\ell}_{ij}}{\ell_{ij}} \frac{\ell_{jk}}{\tilde{\ell}_{jk}} \frac{\tilde{\ell}_{ki}}{\ell_{ki}}, \tag{5.2}$$

and similarly for u_j, u_k. □

At this point it might be tempting to believe that all discrete metrics are conformally equivalent, since they are equivalent on individual triangles. However, since scale factors (at vertices) are shared by several triangles, one still obtains an appropriate degree of rigidity. In particular, an important observation is that all metrics within the same discrete conformal equivalence class can be identified with a canonical collection of *length cross ratios* [**SSP08**]:

DEFINITION. Let $\ell : \mathsf{E} \to \mathbb{R}_{>0}$ be a discrete metric on a triangulated surface $\mathsf{M} = (\mathsf{V}, \mathsf{E}, \mathsf{F})$. For any pair of triangles $ijk, jil \in \mathsf{F}$ sharing a common edge $ij \in \mathsf{E}$, the associated *length cross ratio* is the quantity

$$\mathfrak{c}_{ij} := \frac{\ell_{il}\ell_{jk}}{\ell_{ki}\ell_{lj}}. \tag{5.3}$$

THEOREM 5.3. *Two discrete metrics $\ell, \tilde{\ell}$ on the same triangulation M are discretely conformally equivalent if and only if they induce the same length cross ratios $\mathfrak{c}, \tilde{\mathfrak{c}}$.*

PROOF. First suppose that $\ell, \tilde{\ell}$ are discretely conformally equivalent, i.e., that $\tilde{\ell}_{ij} = e^{(u_i+u_j)/2}\ell_{ij}$ for some collection of log conformal factors $u_i : \mathsf{V} \to \mathbb{R}$. Then

$$\tilde{\mathfrak{c}}_{ijkl} = \frac{\tilde{\ell}_{il}\tilde{\ell}_{jk}}{\tilde{\ell}_{ki}\tilde{\ell}_{lj}} = \frac{e^{(u_i+u_l)/2}e^{(u_j+u_k)/2}\ell_{il}\ell_{jk}}{e^{(u_k+u_i)/2}e^{(u_l+u_j)/2}\ell_{ki}\ell_{lj}} = \frac{\ell_{il}\ell_{jk}}{\ell_{ki}\ell_{lj}} = \mathfrak{c}_{ijkl}.$$

Now suppose that $\ell, \tilde{\ell}$ induce identical cross ratios, i.e., $\mathfrak{c} = \tilde{\mathfrak{c}}$. By Lemma 5.2, these metrics already satisfy the conformal equivalence relation on individual triangles. In particular, for any pair of adjacent triangles ijk, jil, we can find compatible log scale factors u_i, u_j, u_k and v_j, v_i, v_l. Moreover, these scale factors will agree on the shared edge ij if and only if $u_i = v_i$ and $u_j = v_j$. Applying Equation (5.2) we discover that this compatibility condition is equivalent to

$$\frac{\ell_{jk}}{\tilde{\ell}_{jk}}\frac{\tilde{\ell}_{ki}}{\ell_{ki}} = \frac{\tilde{\ell}_{il}}{\ell_{il}}\frac{\ell_{lj}}{\tilde{\ell}_{lj}},$$

i.e., equality of cross ratios. □

The length cross ratios $\mathfrak{c} : E \to \mathbb{R}$ effectively specify a point in the Teichmüller space of discrete metrics; see [**BPS15**, Remark 2.1.2] for further discussion. Extrinsically, it is not hard to show that for a discrete surface immersed in $\mathbb{R}^n$, length cross ratios will be preserved by Möbius transformations of the vertices (assuming that the transformed vertices are connected by straight segments rather than circular arcs). As in the smooth setting, however, the space of maps $\mathsf{f} : \mathsf{V} \to \mathbb{C}$ that induce a discretely conformally equivalent metric $\tilde{\ell}_{ij} := |\mathsf{f}_j - \mathsf{f}_i|$ is much larger than just the space of Möbius transformations. One characterization is that discrete conformal maps (in this sense) correspond to piecewise projective maps that preserve triangle circumcircles [**SSP08**, Section 3.4]. Connections between conformally equivalent discrete metrics and piecewise Möbius transformations have also been studied [**VMW15**].

Two elementary examples help reinforce the connection between smooth and discrete conformal equivalence:

EXAMPLE 5.4 (discrete metrics on the two-triangle sphere). In the smooth setting, all 2-spheres have the same conformal structure. Consider the triangulation of the 2-sphere by two triangles depicted in Figure 2, bottom. Any discrete metric on this triangulation is determined by three edge lengths $\ell_{ij}, \ell_{jk}, \ell_{ki} > 0$ satisfying the triangle inequality. By the same argument as in Lemma 5.2, all discrete metrics on this triangulation of the sphere are discretely conformally equivalent—mirroring the situation in the smooth setting.

EXAMPLE 5.5 (discrete metrics on the two-triangle torus). In the smooth setting, there is a two-parameter family of conformal structures on the torus. Consider the triangulation of the torus depicted in Figure 2, top. Any discrete metric on this triangulation is determined by three edge lengths a, b, c corresponding to the edge marked with a single arrow, the edge marked with a double arrow, and the diagonal. Since there is only one vertex in the triangulation (and hence one scale factor), two such metrics are discretely conformally equivalent if and only if they are related by a uniform scaling—partitioning the three-parameter family of discrete metrics into a two-parameter family of conformal classes (as in the smooth setting).

There are however major difficulties with the theory discussed so far: it only allows us to discuss conformal equivalence of polyhedra with identical combinatorics, since it depends on comparing values associated with edges—at this point, for instance, it is not even possible to say that two different triangulations of the cube (with square faces split along different diagonals) are discretely conformally equivalent, even though they are isometric in the usual sense. Moreover, procedures for uniformizing a given discrete metric (discussed in Section 5.3) may break down before achieving constant curvature, i.e., the metric may degenerate into a collection of edge lengths where the triangle inequality no longer holds. In order to develop a full theory of discrete uniformization, we therefore have to understand what it means for two combinatorially *inequivalent* polyhedra to be conformally equivalent. For this, we first need to consider a structure called the *intrinsic Delaunay triangulation*, which enables us to associate a canonical triangulation with any polyhedral cone metric.

5.2. Intrinsic delaunay triangulations. *Delaunay triangulations* arise naturally throughout discrete and computational geometry, and have numerous characterizations. For a triangulation in the plane, the most typical characterization is that the circumcenter of every triangle is "empty," i.e., no vertices of the triangulation are contained in its interior. An equivalent condition is that the sum of angles opposite every interior edge must be no greater than *pi*, i.e., for any two triangles ijk, jil sharing an edge jil,

$$\theta_k^{ij} + \theta_l^{ji} \leq \pi. \tag{5.4}$$

Unlike the empty circumcircle definition, the angle sum condition generalizes in a straightforward way to any triangulated surface $\mathsf{M} = (\mathsf{V}, \mathsf{E}, \mathsf{F})$ with discrete metric $\ell : \mathsf{E} \to \mathbb{R}_{>0}$, since interior angles θ are determined (via the law of sines) by the edge lengths ℓ. Any discrete surface (M, ℓ) satisfying Equation (5.4) at each edge is called *intrinsic Delaunay.*

Every polyhedral cone metric admits a Delaunay triangulation, i.e., given a surface M with cone metric g, there always exists at least one triangulation of the cone points (vertices) where all edges satisfy Equation (5.4) [**ILTC01**]. In an extrinsic setting, one can imagine that a given polyhedron in $\mathbb{R}^3$ is triangulated by different collections of geodesic arcs terminating at vertices—not necessarily corresponding to the extrinsic edges (see especially Figure 15, right). Each such triangulation defines a discrete metric (given by the geodesic edge lengths), and one such metric will always be Delaunay. A subtlety here is that this triangulation will not in general be simplicial—instead, one must consider *irregular* triangulations where, for example, two edges of the same triangle might be identified. (Such triangulations can be expressed as Δ-complexes, as discussed in Section 2).

As long as a Delaunay triangulation has no cocircular pairs of triangles, then it is also unique [**BS07**]. In other words, if two adjacent triangles ijk, jil share a common circumcircle when laid out in the plane, then either triangulation of the quadrilateral $ijkl$ will satisfy Equation (5.4) with equality (since opposite angles in a circular quadrilateral always sum to π). In general, replacing one triangulation of a quad $ijkl$ with the other is referred to as an *edge flip.* Given an initial triangulation, an intrinsic Delaunay triangulation can be obtained via a simple algorithm: flip non-Delaunay edges (in any order) until they all satisfy Equation (5.4); this algorithm will always terminate in a finite number of flips [**BS07**].

5.3. Discrete uniformization. With several key pieces in place, one can now define a notion of discrete conformal equivalence for polyhedral surfaces with different combinatorics; this definition in turn leads naturally to a discrete uniformization

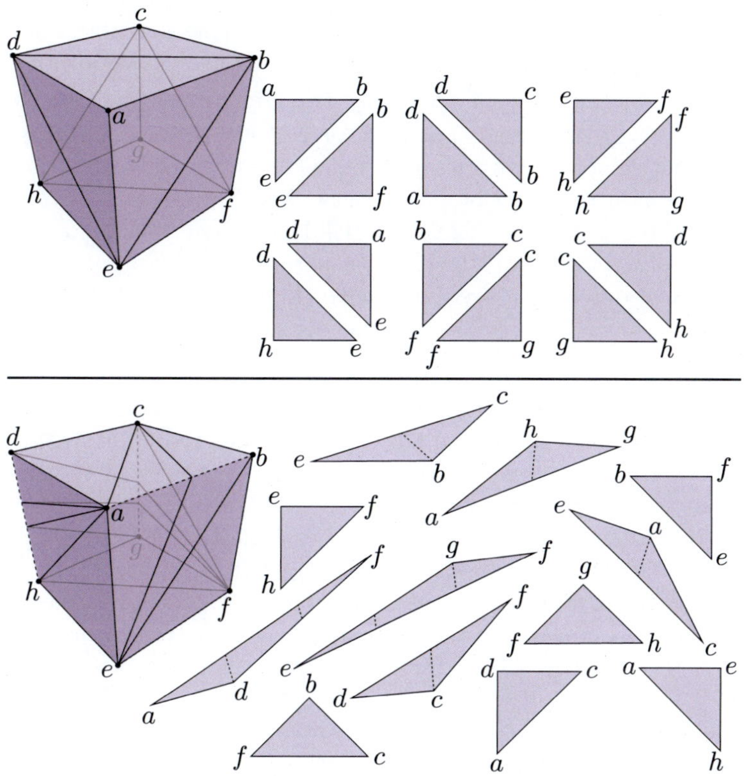

FIGURE 15. A given polyhedron can be triangulated in many different ways; from an intrinsic point of view it does not matter whether there is a simplicial embedding into $\mathbb{R}^n$. For instance, the triangulations depicted at top and bottom both describe the same cone metric.

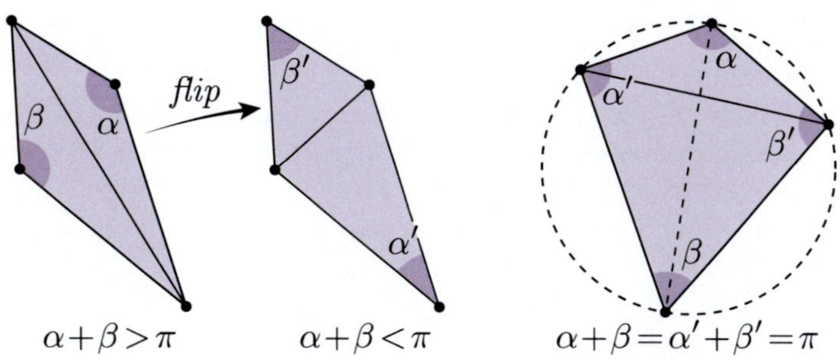

FIGURE 16. Left: an edge is Delaunay if the sum of opposite angles α, β is no greater than π; "flipping" a non-Delaunay edge always yields a Delaunay configuration. Right: the angle sum condition holds with equality if and only if two triangles are cocircular.

theorem. There are two equivalent ways to state this definition, one which appeals only to Euclidean geometry, and another that makes use of a hyperbolic metric naturally associated with any Euclidean polyhedron (to be discussed in Part IV):

DEFINITION 5.6. Let (M, ℓ) and $(\widetilde{\mathsf{M}}, \tilde{\ell})$ be Delaunay triangulations of the same topological surface and with the same vertex set V. These triangulations are **discretely conformally equivalent** if either of two equivalent conditions hold:

I. There is a sequence of Delaunay triangulations $(\mathsf{M}, \ell) = (\mathsf{M}_1, \ell_1), \ldots, (\mathsf{M}_n, \ell_n) = (\widetilde{\mathsf{M}}, \tilde{\ell})$ such that each pair of consecutive surfaces has either (i) identical combinatorics and equal length cross ratios (i.e., they are related by a rescaling of edge lengths à la Equation (5.1)), or (ii) different combinatorics but identical Euclidean metric (i.e., they are related by performing Euclidean edge flips on pairs of cocircular triangles).
II. The ideal polyhedra associated with (M, ℓ) and $(\widetilde{\mathsf{M}}, \tilde{\ell})$ (as defined in Section 6.2) are related by a hyperbolic isometry.

This definition might seem somewhat limited, since at first glance it seems to apply "only" to Delaunay triangulations. Yet since every Euclidean polyhedron has a Delaunay triangulation, Definition 5.6 is in fact much more natural than the one discussed in Section 5.1: it places emphasis on the geometry of the surface itself, rather than an arbitrary triangulation that merely sits on top of the surface.

Discrete uniformization then amounts to essentially the same statement as in the smooth setting: given any (discrete) metric, there is a (discretely) conformally equivalent one with constant curvature. More precisely, for a given discrete surface (M, ℓ), there exists a conformally equivalent discrete metric with either constant cone angle Ω_i (Equation (2.1)), or more generally, that achieves some prescribed cone angle Ω_i^* (so long as it satisfies the discrete analogue of Gauss-Bonnet mentioned in Section 2). Several closely related versions of discrete uniformization are encapsulated by the following theorems.

THEOREM 5.7 (spherical uniformization). *For any closed finite discrete surface (M, ℓ) of genus zero, there exists a discretely conformally equivalent convex polyhedron $(\widetilde{\mathsf{M}}, \tilde{\ell})$ inscribed in the unit sphere $S^2 \subset \mathbb{R}^3$.*

THEOREM 5.8 (Euclidean uniformization). *For any closed finite genus-1 discrete surface (M, ℓ) with vertex set V, there exists a discretely conformally equivalent surface $(\widetilde{\mathsf{M}}, \tilde{\ell})$ with zero curvature at each vertex, i.e., $\Omega_i^* = 0$ for all $i \in \mathsf{V}$.*

In the hyperbolic case (i.e., for genus $g \geq 2$), uniformization must be treated in a different way, since unlike the spherical case we cannot isometrically embed H^2 into $\mathbb{R}^3$. Instead we endow our triangulation with a piecewise hyperbolic metric, this time constructed from ordinary rather than ideal hyperbolic triangles. In this setting, two discrete metrics $\ell, \tilde{\ell} : \mathsf{E} \to \mathbb{R}_{>0}$ on the same triangulation $\mathsf{M} = (\mathsf{V}, \mathsf{E}, \mathsf{F})$ are discretely conformally equivalent if they satisfy the relationship

$$\sinh(\tilde{\ell}_{ij}/2) = e^{(u_i+u_j)/2} \sinh(\ell_{ij}/2)$$

for each edge $ij \in \mathsf{E}$ (for further discussion see [**BPS15**]). The definition of the cone angle Ω_i is unchanged: it is the difference between 2π and the angle sum Θ_i of the interior angles at vertex i. (This data also describes a Euclidean polyhedron inscribed in the hyperboloid model, where the Euclidean length of each edge is determined by the indefinite inner product $\langle x, x \rangle := x_1^2 + x_2^2 - x_3^2$.) The definition of conformal equivalence for variable triangulations is then directly analogous to Definition 5.6, yielding the following uniformization theorem:

THEOREM 5.9 (hyperbolic uniformization). *For any closed finite discrete surface* (M, ℓ) *of genus* $g \geq 2$ *with a piecewise hyperbolic metric, there exists a discretely conformally equivalent piecewise hyperbolic surface* $(\widetilde{\mathsf{M}}, \tilde{\ell})$ *with zero curvature at every vertex* ($\Omega_i = 0$ *for all* $i \in \mathsf{V}$).

Theorem 5.8 is a special case of a more general uniformization theorem which guarantees existence of metrics with prescribed cone angles at each vertex:

THEOREM 5.10. *For any closed finite genus-g discrete surface* (M, ℓ) *with vertex set* V, *and any set of target cone angles* $\Omega^* : \mathsf{V} \to (-\infty, 2\pi)$ *satisfying the Gauss-Bonnet condition*

$$\sum_{i \in \mathsf{V}} \Omega_i^* = 2\pi(2 - 2g),$$

there exists a discretely conformally equivalent surface $(\widetilde{\mathsf{M}}, \tilde{\ell})$ *with* $\widetilde{\Omega}_i = \Omega_i^*$.

From the Euclidean point of view, the basic idea behind proving discrete uniformization is to consider a time-continuous evolution of the discrete metric, akin to the same smooth Ricci/Yamabe flow discussed in Section 4.3. This flow was first studied by Luo in the context of a fixed triangulation [**Luo04**]. In this setting, the scale factors $u : \mathsf{V} \to \mathbb{R}$ evolve according to the flow

$$\tfrac{d}{dt} u_i(t) = \Omega_i(t) - \Omega_i^*, \tag{5.5}$$

where $\Omega_i(t)$ is the angle defect corresponding to the current edge lengths $\tilde{\ell}(t) := e^{(u_i(t)+u_j(t))/2} \ell_{ij}$, and Ω_i^* is the (fixed) target curvature. Note that the curvatures $\Omega(t)$ depend on the angles θ, which in turn depend on the rescaled edge lengths $\ell_{ij}(t) = e^{(u_i(t)+u_j(t))/2} \ell_{ij}(0)$. Differentiating Equation (5.5) with respect to u reveals that the flow is the gradient flow on a convex energy $E(u)$, whose Hessian is given explicitly by the cotangent discretization of the Laplace-Beltrami operator Δ [**Luo04**]. An explicit form for this energy was first given by Springborn et al. [**SSP08**], and will be discussed in Section 6.4. Due to the convexity of the energy, the flow yields a uniformizing collection of scale factors $u^* : \mathsf{V} \to \mathbb{R}$ in finite time—*if* such factors exist. However, for a fixed triangulation the flow may become singular, i.e., it may reach a point where the edge lengths no longer describe a valid discrete metric (due to failure of the triangle inequality).

For this reason one must in general define discrete uniformization in terms of variable triangulations. Here one considers the same evolution of scale factors $u_i(t)$ as in Equation (5.5), but where the corresponding discrete metric $\tilde{\ell}(t)$ is always defined with respect to the intrinsic Delaunay triangulation of the current cone metric. At certain critical moments (namely, when there is a cocircular pair of triangles) this triangulation will not be unique, and effectively experiences an edge flip as the flow continues to evolve. One can continue to think of this flow as a gradient flow on a piecewise smooth energy whose definition is uniform over *Penner cells*, i.e., regions of $\mathbb{R}^{|\mathsf{V}|}$ over which the intrinsic Delaunay triangulation induced by the scale factors $u : V \to \mathbb{R}$ does not change. This energy remains C^1 (in fact C^2) everywhere—even at the boundary of such cells [**Spr17**]. One therefore has a well-defined gradient flow on a convex energy, and can then show that any discrete metric (with no special conditions on geometry or combinatorics) will flow to one of constant curvature in finite time.

The Euclidean case (Theorem 5.8 and Theorem 5.10), was first proven by Gu et al. [**GLSW18**]; the hyperbolic case (Theorem 5.9) is implied by a proof by Fillastre [**Fil08**] about hyperbolic polyhedra, and was shown more directly by Gu et al. [**GGL$^+$18**]. The spherical case (Theorem 5.7) was first shown by Springborn [**Spr17**], via a proof that also covers the Euclidean and hyperbolic cases. None of these proofs consider domains with boundary, nor conditions on monodromy around noncontractible cycles [**CZ17**]. In order for a uniformization to be algorithmically computable, it may also be important (depending on the choice of algorithm) that the number of edge flips encountered during the flow is finite, i.e., that the flow passes through only finitely many Penner cells; this fact was established by Wu [**Wu14**].

Convergence under mesh refinement of discrete conformal maps to smooth ones has been studied both numerically and analytically [**SWGL15**, **Büc16**, **Büc18a**] (see also [**SC17**, Figure 16]), though many questions still remain.

6. Part IV: Hyperbolic polyhedra

In this final part we consider the hyperbolic viewpoint on discrete conformal geometry, which helps to simplify and unify the story. Along the way we will also encounter some fascinating connections between geometry and combinatorics, which instigated the development of techniques now used for discrete conformal geometry.

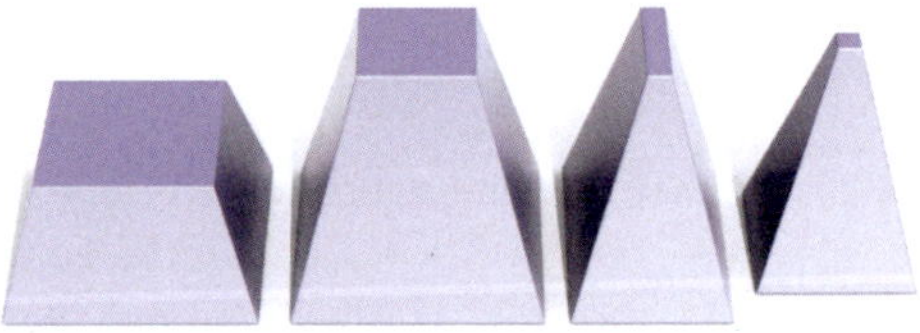

Why is hyperbolic geometry an effective framework for problems in discrete conformal geometry? For one thing, ideal hyperbolic polyhedra are often "easier" to construct than Euclidean ones, since they typically exhibit greater rigidity. For instance, the dihedral angles of a convex polyhedron do not determine its extrinsic geometry in the Euclidean case (consider, for instance, a family of truncated polyhedra), whereas convex hyperbolic polyhedra are essentially determined by their dihedral angles [**HR93**]. Moreover, even though Alexandrov's theorem guarantees that a convex Euclidean polyhedron is uniquely determined by its metric (up to rigid motions), actually constructing an embedding is quite challenging [**IBI08**, **KPD09**]. In contrast, an embedding of an ideal convex hyperbolic polyhedron with prescribed metric can be obtained as the minimizer of a convex energy [**Spr17**]. These two realization problems—ideal polyhedra with prescribed dihedral angles or prescribed metric—turn out to be closely connected (and in some cases, identical) to the problem of finding circle patterns with prescribed intersection angles, or discretely conformally equivalent metrics with prescribed cone angles, resp.

In order to describe these connections, we first give some background on hyperbolic geometry, followed by a discussion of variational principles for hyperbolic polyhedra and their connection to discrete conformal maps.

6.1. Hyperbolic geometry. The hyperbolic plane H^2 is topologically like the ordinary Euclidean plane, but has "saddle-like" geometry everywhere; more formally, it is a complete and simply-connected surface of constant negative curvature $K = -1$. Unlike the unit 2-sphere ($K = +1$), the hyperbolic plane cannot be isometrically embedded in $\mathbb{R}^3$ and must instead be studied in terms of various

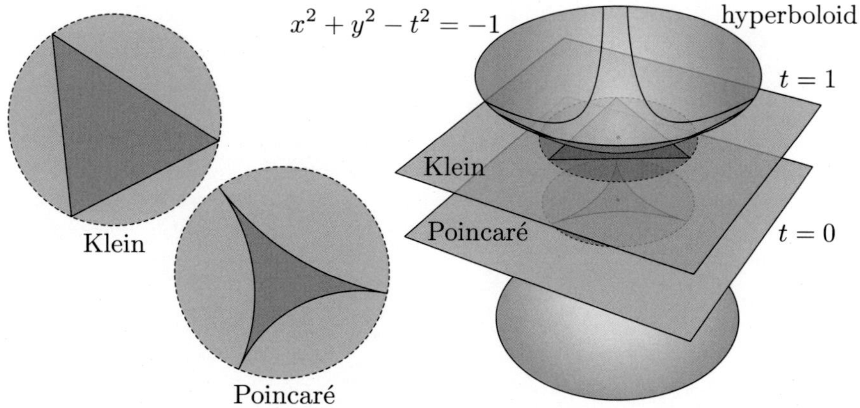

FIGURE 17. Left: an ideal triangle in two models of the hyperbolic plane H^2, which preserve either straight lines (Klein) or angles (Poincaré). Right: these models are both related to the hyperboloid model via projections onto the planes $t = 0$ and $t = 1$ (resp.).

models. Two models that naturally arise in discrete conformal geometry are the *Poincaré disk model* and the *Klein disk model*, which both model H^2 on the open unit disk $D^2 \subset \mathbb{R}^2$, as depicted in Figure 17, *left.* The most salient geometric facts about these two models is that the Poincaré model is conformal and hence faithfully represents angles but not straight lines (geodesics), whereas the Klein model faithfully represents geodesics as straight lines but does not preserve angles. In the Poincaré model, geodesics instead become circular arcs orthogonal to the unit circle ∂D^2. These two models can be connected via a third *hyperboloid model*, which puts H^2 into correspondence with one sheet of the hyperboloid $x^2+y^2-t^2 = -1$; in this model, geodesics are represented as intersections of the hyperboloid with planes through the origin. If we express points on the hyperboloid in homogeneous coordinates (x, y, t), then the map from the hyperboloid to the Klein model is given by the usual homogeneous projection $(x, y, t) \mapsto (x/t, y/t)$. Similarly, to get from the hyperboloid model to the Poincaré model we can simply shift the hyperboloid a unit distance along the t-axis before applying the same projection. These relationships are depicted in Figure 17, *right.*

Unlike the Euclidean plane, where a pair of parallel lines remains at a constant distance, one can construct pairs of hyperbolic geodesics that never intersect yet become arbitrarily close as they approach infinity (such pairs are sometimes called *limiting parallels*). An *ideal hyperbolic triangle* is a figure bounded by three such pairs; its *vertices* are the three corresponding limit points on the sphere at infinity. As a result, all interior angles are zero and all edges have infinite length. In fact, all ideal triangles are *congruent*, i.e., they are identical up to isometries of H^2, which are represented in the Poincaré model and the Klein model as Möbius transformations and projective transformations (resp.) that map the disk to itself.

6.2. Ideal hyperbolic polyhedra. Any Euclidean triangulation can naturally be identified with one made of ideal hyperbolic triangles: take each triangle and draw it in the Euclidean plane. The circumcircle of this triangle (i.e., the

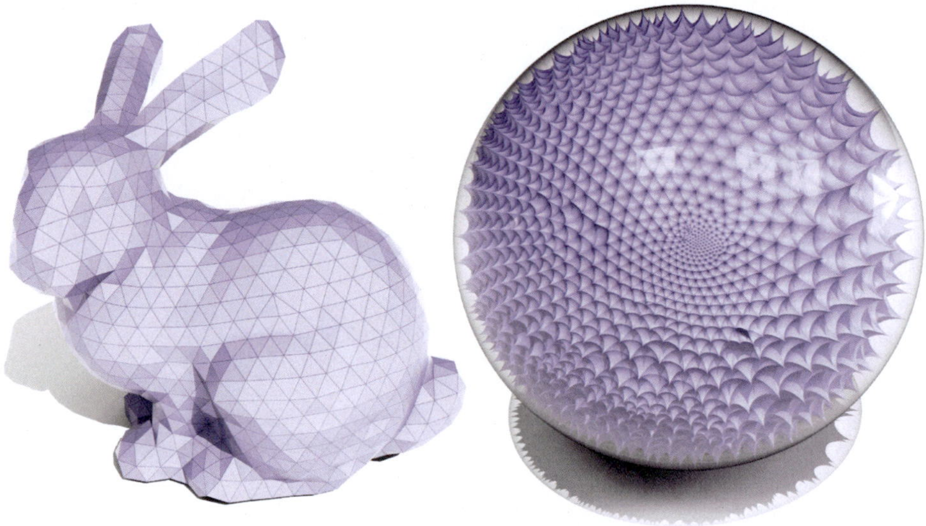

FIGURE 18. A Euclidean polyhedron and its realization as an ideal hyperbolic polyhedron.

unique circle passing through its three vertices) can be viewed as a copy of the hyperbolic plane in the Klein model. The three straight edges then correspond to three hyperbolic geodesics, and the triangle itself becomes an ideal hyperbolic triangle. By gluing these ideal triangles together along shared edges, we obtain an *ideal hyperbolic polyhedron*, i.e., a surface of constant curvature $K = 1$ away from a collection of *cusps* at vertices, which extend to infinity (examples are shown in Figure 18 and Figure 22).

Since all ideal triangles are congruent, it would be easy to think that the geometry of such a polyhedron is determined purely by the combinatorics of M. However, additional structure is encoded in the way pairs of ideal triangles are identified: given two ideal triangles ijk, jil sharing an edge γ, we can apply a hyperbolic isometry to one of them (say, jil) that fixes γ, i.e., we can "slide" one ideal triangle along the other. The intrinsic geometry of the hyperbolic polyhedron is therefore fully determined by (i) the combinatorics of the triangulation, and (ii) data that specifies how edges are identified.

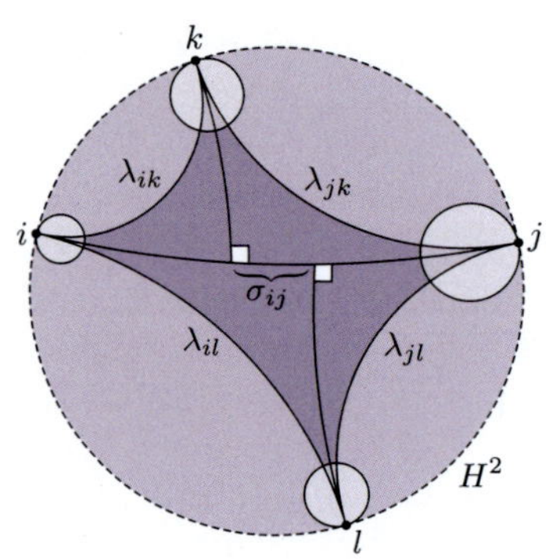

To explicitly encode identifications, Penner [**Pen87**] defines *shear coordinates* $\sigma_{ij} \in \mathbb{R}$ per edge $ij \in \mathsf{E}$ which measure the distance between the altitudes of the two triangles with base ij, as depicted in the inset. (Note that these distances are measured with respect to the hyperbolic metric, and have nothing to do with a particular model or choice of coordinates on H^2.) Alternatively, one can "decorate" the vertices of the ideal polyhedron with arbitrarily-chosen *horocycles*, which are copies of Euclidean $\mathbb{R}^n$ sitting inside H^n. Just as a Euclidean line can be viewed as a circle of infinite radius, a horocycle can be viewed as the limit of a family of mutually tangent hyperbolic circles as the radius goes to infinity—in the Poincaré model, any such curve is represented by a circle tangent to the boundary. For a triangle ijk, the horocycle associated with a given vertex i will meet the two hyperbolic geodesics corresponding to edges ij and ik orthogonally. The *Penner coordinate* $\lambda_{ij} \in \mathbb{R}$ of each edge ij is then the (hyperbolic) distance between the horocycles at i and j. More precisely, λ_{ij} gives the *signed* distance between horocycles: positive if they are disjoint, negative if they intersect. Since the choice of horocycles is arbitrary, the Penner coordinate of a *single* edge does not, in isolation, provide any geometric information about the hyperbolic structure of the polyhedron. However, the Penner coordinates share an important relationship with the shear coordinates, namely

$$2\sigma_{ij} = \lambda_{jk} - \lambda_{ik} + \lambda_{il} - \lambda_{jl}. \tag{6.1}$$

Hence, two sets of Penner coordinates determine the same ideal hyperbolic polyhedron if and only if they induce the same shear coordinates.

The similarity between Equation (6.1) and the definition of the length cross ratio (Equation (5.3)) is not superficial: when an ideal hyperbolic polyhedron is constructed from a Euclidean polyhedron (as described above), the shear coordinates and length cross ratios will be related by

$$\sigma_{ij} = \log(\mathfrak{c}_{ij}),$$

independent of any choice of horocycles. It is therefore natural to pick horocycles such that

$$\lambda_{ij} = 2\log \ell_{ij},$$

i.e., such that the Penner coordinates encode the edge lengths. From the conformal point of view, the shear coordinates encode the discrete conformal equivalence class, and the Penner coordinates specify a particular discrete metric within this class.

Importantly, since both the shear coordinates and Penner coordinates are hyperbolic distances, they will be preserved by hyperbolic isometries. Hence, *isometry classes of ideal hyperbolic polyhedra correspond to conformal equivalence classes of discrete metrics.* This basic viewpoint was introduced by Bobenko, Pinkall, and Springborn; see in particular [**BPS15**, Section 5] for further discussion.

6.3. Variational principles. A *variational principle* expresses the solution to a problem as a minimizer of an energy functional (possibly subject to constraints). A major appeal of discrete conformal geometry is that the variational principles involved are often *convex*, providing a clear picture of existence and uniqueness, and allowing one to leverage principled methods from convex optimization [**BV04**] to develop practical algorithms with clear guarantees. Much of the understanding of these principles originated in the hyperbolic setting, where they can be given a concrete geometric interpretation. Here, the basic question is not how to uniformize a Euclidean polyhedron, but rather how to construct embeddings of hyperbolic polyhedra with prescribed geometric data—namely, dihedral angles or the intrinsic metric.

FIGURE 19. Unlike the octahedron (top), the stellated octahedron (bottom) cannot be realized a convex spherical polyhedron, no matter where the vertices are placed.

A Prelude: Steiner's Problem. Why is anything known about hyperbolic polyhedra in the first place? One answer is that triangulations of hyperbolic space are a key tool for understanding the geometry and topology of 3-manifolds [**Thu98**]. A more basic question was posed by the 19th century Swiss geometer Jakob Steiner, which provides a first glimpse at some of the deep connections between geometry and combinatorics:

> *"Wenn irgend ein convexes Polyëder gegeben ist, läßt sich dann immer (oder in welchen Fällen nur) irgend ein anderes, welches mit ihm in Hinsicht der Art und der Zusammensetzung der Grenzflächen übereinstimmt (oder von gleicher Gattung ist), in oder um eine Kugelfläche, oder in oder um irgend eine andere Fläche zweiten Grades beschreiben (d. h. daß seine Ecken alle in dieser Fläche liegen, oder seine Grenzflächen alle diese Fläche berühren)?"*

In English: *"Given a convex polyhedron, when can you find a combinatorially equivalent convex polyhedron that is inscribed in the sphere (or another quadratic surface)?"* More abstractly, we might ask: which combinatorial tessellations of the 2-sphere can be realized as convex polyhedra with vertices on the sphere? Steiner's question is very much in the spirit of discrete differential geometry: he is looking for a finite analogue of the sphere that exactly preserves a key property of its smooth counterpart (convexity).

Surprisingly enough, not all tessellations of the sphere can be realized as convex spherical polyhedra: for instance the octahedron sits in the sphere, but for the *stellated octahedron* (where we split each face into three) there turns out to be no way to arrange the vertices on the sphere so that it becomes convex [**Riv93**, Corollary 10]. In other words, only certain "nice" combinatorial tessellations of the topological sphere can be interpreted as geometric spheres. This same question can also be stated in terms of Delaunay triangulations: every planar Delaunay triangulation is the image of some sphere-inscribed convex polyhedron under stereographic projection. Hence, Steiner's question amounts to asking which combinatorial triangulations can be realized as planar Delaunay triangulations.

More than 150 years later, a general solution to Steiner's problem was found by framing it as a problem about hyperbolic rather than Euclidean polyhedra [**Riv96**]. The bridge is the one we have already seen (in Section 6.2): Euclidean polyhedra determine a hyperbolic structure by interpreting each Euclidean triangle as an ideal triangle in the Klein model; constructing a convex sphere-inscribed Euclidean polyhedron with prescribed combinatorics then becomes equivalent to finding a convex ideal hyperbolic polyhedron with the same combinatorics. As we will see, this problem turns out to be closely linked to the variational principles for both the circle patterns studied in Part II, and the conformally equivalent discrete metrics studied in Part III.

Lobachevsky's Function. Perhaps the most concrete link between hyperbolic and discrete conformal geometry is through the *Lobachevsky function*

$$(\theta) := -\int_0^\theta \log|2\sin u|\,du,$$

Thurston provides an account of some of its analytical and geometric properties [**Thu79**, Chapter 7]. Though originally developed for calculating hyperbolic volumes, this function also arises naturally in variational principles for discrete conformal maps.

In particular, consider an *ideal hyperbolic tetrahedron*, i.e., an ideal hyperbolic polyhedron with the usual combinatorics of a tetrahedron. One can show that the

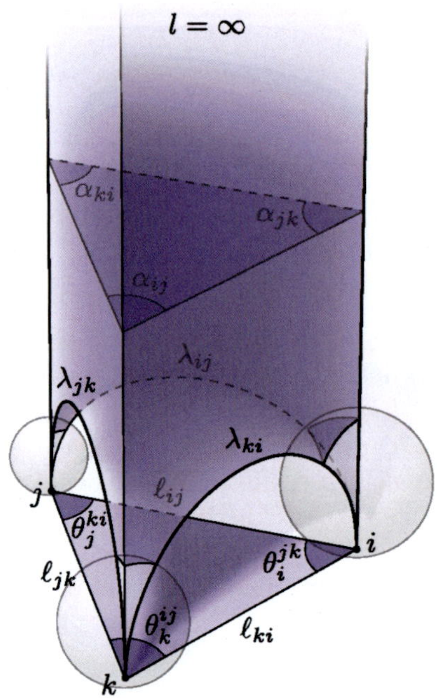

FIGURE 20. Any Euclidean triangle ijk with interior angles θ can be identified with an ideal hyperbolic tetrahedron $ijkl$ with identical dihedral angles. The logarithmic edge lengths $\lambda_{ij} = 2 \log \ell_{ij}$ determine a choice of horospheres at vertices.

volume of such a tetrahedron is

$$V(\alpha) := \tfrac{1}{2} \sum_{ij} (\alpha_{ij}),$$

where the sum is taken over all six edges, and α_{ij} is the dihedral angle associated with edge ij (for a proof, see [**Thu79**, Theorem 7.2.1]). Since opposite dihedral angles in an ideal tetrahedron are equal, only three of the dihedral angles are distinct. Moreover, these angles, which we will call α_{ij}, α_{jk}, and α_{ki}, can be naturally identified with the interior angles $\theta_k^{ij}, \theta_i^{jk}, \theta_j^{ki}$ of a Euclidean triangle (resp.), as depicted in Figure 20. Importantly, then, the function $-V$ is convex when restricted to the set of valid Euclidean angles, i.e., whenever $\alpha_{ij} + \alpha_{jk} + \alpha_{ki} = \pi$ and all the angles are positive [**Riv94**, Theorem 2.1]. The first order change in volume $\dot{V}$ with respect to a first order change $\dot{\alpha}$ in dihedral angles can be expressed via the following *Schläfli formula* for ideal tetrahedra:

$$\dot{V} = -\frac{1}{2} \sum_{ij} \lambda_{ij} \dot{\alpha}_{ij}, \tag{6.2}$$

where λ_{ij} are the Penner coordinates of edge ij [**Mil94**]; Rivin provides a proof [**Riv94**, Theorem 14.5].

Such observations provide a starting point for convex variational principles for hyperbolic polyhedra [**HR93, Riv93, Riv96, Fil08**], which in turn provide variational principles for discrete conformal geometry. Consider, for instance, a Euclidean circle pattern associated with the faces of a triangulation M, where circles

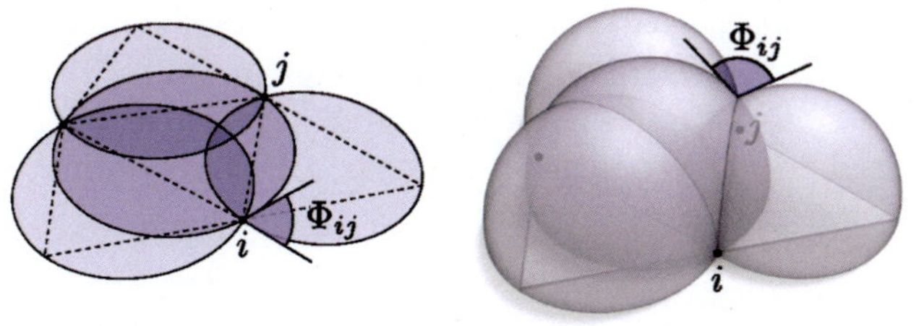

FIGURE 21. In the Poincaré halfspace model for H^3, two Euclidean circles intersecting at an angle Φ_{ij} can be interpreted as two copies of H^2 intersecting at the same dihedral angle. Three such intersections trace out an ideal hyperbolic triangle.

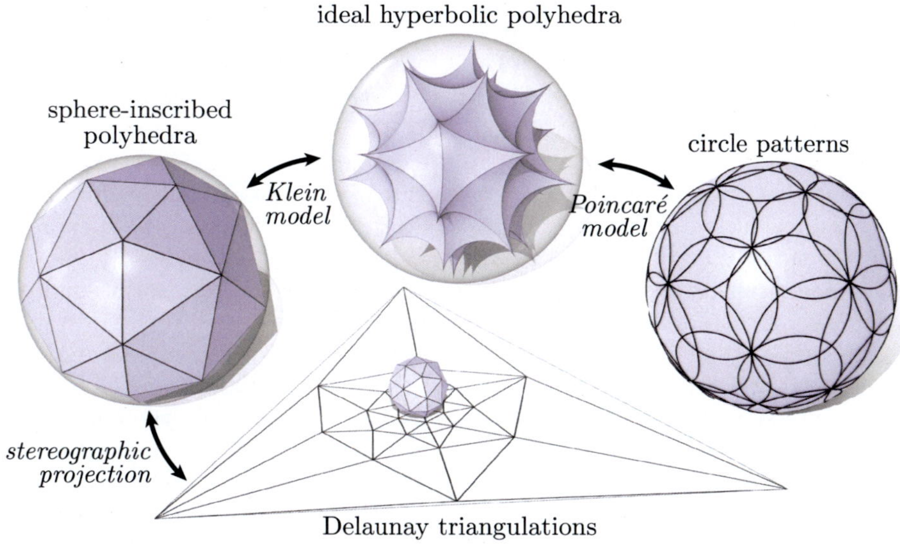

FIGURE 22. A triangulation with fixed combinatorics can be realized as a Delaunay triangulation in the plane if and only if it can also be realized as a convex Euclidean polyhedron inscribed in the sphere, a convex ideal hyperbolic triangulation, and a circle pattern on the sphere (with circles associated with triangles).

associated with neighboring triangles ijk, jil intersect at angles Φ_{ij} [**BS04a**]. If we view each circle as the boundary of a hemisphere sitting on top of the plane, then the intersection angles coincide with the dihedral angles between neighboring spheres. In the Poincaré half space model for H^3, each such hemisphere corresponds to a copy of H^2. The intersections between spheres are then hyperbolic geodesics bounding ideal hyperbolic triangles. Hence, circle patterns with prescribed intersection angles are equivalent to convex ideal polyhedra with prescribed dihedral angles, which are in turn equivalent to convex Euclidean polyhedra inscribed in the sphere (via the Klein model, as discussed in Section 6.2), which are in turn equivalent to planar Delaunay triangulations, via stereographic projection. The connections between these different points of view are summarized in Figure 22.

Alternatively, suppose we draw an ideal tetrahedron in the Poincaré half space model, as depicted in Figure 20. Then the three distinct dihedral angles α_{ij} correspond to the three interior angles of a Euclidean triangle (at infinity), and we can always pick horospheres at vertices such that the Penner coordinates λ_{ij} are determined by the logarithmic edge lengths, i.e., such that $\ell_{ij} = e^{\lambda_{ij}/2}$. By Equation (6.2), the first order variation of the function

$$\varphi(\alpha, x) := V(\alpha) + \sum_{ij} \lambda_{ij}\alpha_{ij} \tag{6.3}$$

is then given by

$$\dot{\varphi} = \dot{V} + \sum_{ij} \dot{\lambda}_{ij}\alpha_{ij} + \sum_{ij} \lambda_{ij}\dot{\alpha}_{ij} = \sum_{ij} \dot{\lambda}_{ij}\alpha_{ij}.$$

Hence,

$$\frac{\partial \varphi}{\partial \lambda_{ij}} = \alpha_{ij},$$

allowing us to obtain the interior angles of a triangle as the derivatives of a convex functional φ (see also [**BPS15**, Section 4.2]). This functional therefore provides a starting point for variational principles for problems involving an evolution of a discrete metric $\ell = e^{\lambda/2}$. In particular, it leads to a variational principle for discrete uniformization involving the cone angles Ω_i (Section 6.4), which can be expressed in terms of sums of interior angles θ, or equivalently, dihedral angles α.

Chronologically, the presentation in this section is backwards: variational principles in discrete conformal geometry were historically developed by first establishing some kind of "flow" involving only derivatives (such as the combinatorial Ricci flow discussed in Section 4.3); only later were these derivatives integrated to obtain explicit expressions for the underlying *variational principle*, i.e., the energy driving the flow. While derivatives alone are often enough to establish results about existence, uniqueness, convergence, etc., variational principles are quite valuable for understanding the global picture (including connections between Euclidean and hyperbolic problems) as well as for practical algorithms (enabling, for instance, the use of principled optimization strategies [**KSS06**, **SSP08**]).

Circle Patterns. Early approaches to constructing circle packings can be loosely interpreted as iterative algorithms (e.g., Jacobi or Gauss-Seidel) for minimizing an energy [**Thu79**, **CS03**]. Colin de Verdière [**Col91**] gave the first real variational principle for Euclidean and hyperbolic circle patterns with circles at vertices, though only expressions for the derivatives were given; Brägger [**Bra92**] makes a very brief remark that these derivatives appear related to hyperbolic volume (*"Untersucht man diese Funktionale, stösst man rasch auf die Lobachevsky-Funktion und damit auf das Volumen hyperbolischer 3-Simplices."*). Rivin [**Riv94**] considers ideal polyhedra with prescribed dihedral angles, effectively providing a variational principle for Euclidean circle patterns. Bobenko & Springborn [**BS04a**] develop variational principles that connect and generalize the three previously mentioned, and leads to efficient numerical implementation [**KSS06**] due to the absence of difficult constraints. Chow and Luo [**CL03**] subsequently introduce a combinatorial analogue of Ricci flow (Section 4.3), and show that it arises from a convex variational principle by considering the Hessian [**CL03**, Section 3.5]. More recently, Ge also considers a combinatorial analogue of Calabi flow [**Ge18**] suitable for practical computation [**ZLH+18**]. A variety of closely related variational principles and generalizations thereof have also been studied [**Gli11**, **ZGZ$^+$14**, **GX15**], including a large number of applications [**JKG07**].

Conformally Equivalent Discrete Metrics. The first connection between variational principles and conformally equivalent discrete metrics (à la Definition 5.1) came from Roček and Williams' efforts to study quantum gravity through the simplicial calculus of Regge [**Reg61**]. The definition of discrete conformal equivalence here is identical to the one discussed in Section 5.1 [**RW84**, Section VII], though the discrete variational principles are mainly connected to Einstein metrics on four dimensional spacetime rather than metrics of constant scalar curvature on 2-manifolds. Luo [**Luo04**] later developed the same notion of conformally equivalent discrete metrics, and a corresponding discrete Yamabe flow driving the metric toward constant scalar curvature; no explicit energy was given, and existence is guaranteed only if the flow does not develop singularities (i.e., if the discrete metric does not degenerate). Springborn, Schröder, and Pinkall gave an explicit energy in terms of the Lobachevsky function [**SSP08**], and Bobenko, Pinkall, and Springborn [**BPS15**] made the connection to hyperbolic geometry and ideal polyhedra. Through this correspondence, Fillastre effectively proved discrete uniformization in the hyperbolic case [**Fil08**]; Gu et al. establish the same result from the perspective of discrete conformal equivalence [**GGL$^+$18**, Corollary 4 and Theorem 5]. Gu et al. also establish the existence of discrete uniformization in the Euclidean setting [**GLSW18**]; Springborn completes the proof of discrete uniformization via a variational approach that subsumes the Euclidean, hyperbolic, and spherical case [**Spr17**].

Much more has been said about variational principles both for ideal hyperbolic polyhedra, and for polyhedra in general—see for instance [**BS04a**, **DGL08**].

6.4. Variational principle for discrete uniformization. The hyperbolic picture provides an attractive perspective on the discrete uniformization theorems discussed in Section 5.3. Recall that two Euclidean polyhedra are discretely conformally equivalent if and only if their Delaunay triangulations induce the same piecewise hyperbolic metric (Definition 5.6). Intuitively, the reason for using a canonical triangulation like the intrinsic Delaunay triangulation is that different triangulations of the same Euclidean polyhedron will in general induce a *different* hyperbolic metric. The Delaunay triangulation is particularly special because Euclidean and hyperbolic Delaunay flips coincide exactly when the triangulation is not unique, i.e., when two adjacent triangles are cocircular. Consider for instance two triangles $ijk, jil \in \mathsf{F}$, realized as a pair of ideal hyperbolic triangles in the Klein model. In this setting, the new edge length ℓ_{kl} resulting from a flip is determined by *Ptolemy's relation*:

$$\ell_{ij}\ell_{kl} = \ell_{ki}\ell_{lj} + \ell_{il}\ell_{jk}. \tag{6.4}$$

Such a flip is therefore called a *Ptolemy flip*, and preserves the hyperbolic structure of M, i.e., the new and old hyperbolic metric are the same. However, the new and old *Euclidean* metric will be different except in the special case where ijk and jil have the same circumcircle, since in this case the new edge length ℓ_{kl} resulting from a Euclidean flip or a Ptolemy flip will be identical. This fact is somewhat counterintuitive: it says that even for a completely flat domain, a Euclidean edge flip will (in general) change the discrete conformal structure—even though it does not change the Euclidean geometry. Much as a Euclidean edge flip preserves the metric only if the edge is flat (i.e., has zero dihedral angle), a hyperbolic edge flip will preserve the metric only if the hyperbolic edge has zero dihedral angle, or equivalently, if the four vertices are cocyclic. The Delaunay condition is therefore essential for the equivalence of definitions (I) and (II): since all meshes in the sequence are Delaunay, any Euclidean edge flip will also correspond to a Ptolemy flip (and hence preserve the hyperbolic metric). Alternatively one could also say: two triangulations are discretely conformally equivalent if they are related by a sequence of length scalings (à la Equation (5.1)) and *Ptolemy* flips, in which case the intermediate triangulations are no longer required to be Delaunay.

The problem of finding a discretely conformally equivalent flat metric (i.e., all cone angles equal to zero) is then the same as finding an ideal polyhedron in H^3 with prescribed hyperbolic metric. Both problems can be solved by considering the same variational principle, which is discussed at length in a sequence of papers [**SSP08**, **BPS15**, **Spr17**]. In particular, let ℓ be an initial discrete metric on a triangulation M, and for any conformally equivalent metric $\tilde{\ell}$ with scale factors u at vertices, define the following energy with respect to the intrinsic Delaunay triangulation of $\tilde{\ell}$:

$$E(u) := \sum_{ijk \in F} \varphi(\widetilde{\lambda}_{ij}, \widetilde{\lambda}_{jk}, \widetilde{\lambda}_{ki}) - \frac{\pi}{2}(u_i + u_j + u_k) + \frac{1}{2}\sum_{i \in V}(2\pi - \Omega_i^*)u_i.$$

Here, φ is the potential given in Equation (6.3), $\widetilde{\lambda}_{ij} = 2\log(\tilde{\ell}_{ij})$ are the logarithmic edge lengths, and Ω^* are the target cone angles [**SSP08**]. One can show that this energy is convex, and in fact C^2 everywhere [**Spr17**]. To uniformize a given discrete metric over a Euclidean domain (possibly with cone singularities), one can therefore apply standard first- or second-order descent to E with respect to the log conformal factors u; the energy simply needs to be expressed relative to the appropriate intrinsic Delaunay triangulation, which can always be obtained by applying a sequence of Ptolemy flips. This principle (and variations of it) provide one approach to proving the discrete uniformization theorems discussed in Section 5.3 [**Spr17**].

7. Summary

The table below summarizes several different approaches to the discretization of conformal maps. To date, the approach based on conformally equivalent discrete metrics appears to be the only one that clearly exhibits both the right amount of flexibility, *and* furnishes a complete discrete uniformization theorem. However, the perspective of circle preservation comes very close—and ultimately it should not be too surprising to also find a complete picture of discrete uniformization in terms of circles. In fact, although these two perspectives at first appear somewhat disjoint, connections continue to be found at various levels. For instance, Glickenstein [**Gli11**] and Bücking [**Büc18b**] discuss connections between circle patterns and length cross ratios; Lam and Pinkall [**LP16**] show that linearized versions of both theories can be connected to discrete harmonic functions (in the sense of Equation (3.4)). More elementary starting points, natural as they may seem, lead to definitions that turn out to be too rigid or too flexible. However, as noted in Section 1 and discussed throughout, many of these discretizations nonetheless provide interesting connections to the smooth theory, and play an important role in practical algorithms.

Beyond the basic question of flattening/uniformization, many questions about how to discretize conformal geometry remain. For instance, questions about *extrinsic* conformal maps, i.e., transformations of conformal surface immersions in $\mathbb{R}^n$, are relevant for practical tasks in digital geometry processing, where one wishes to directly manipulate geometry sitting in space, rather than mapping it to a canonical domain. Some preliminary work has been done in this direction [**Cra13**], including a recent theory compatible with the notion of conformally equivalent discrete metrics [**LP18**]. There has also been some recent work on applying similar tools to the problem of discrete *geometrization* [**AMY18**], as well as continued interest in problems in numerical relativity [**Gen02**] via Regge calculus.

Approach	Data	Outcome	Comments
ANGLES (Section 3.1)	interior angles θ_i^{jk}	too rigid	similarity of triangles forces single global scale factor
DIRICHLET (Section 3.3)	vertex coordinates f_i	too rigid	(same as ANGLES)
HODGE (Section 3.4)	length ratio w_{ij}	too rigid	uniquely determines a discrete metric
CONJUGATE (Section 3.5)	vertex coordinates f_i	**just right**	*only under refinement / no finite notion of conformal equivalence*
CIRCLES (Section 4)	graph $G = (\mathsf{V}, \mathsf{E})$	too flexible	only combinatorics are considered; no way to distinguish different metrics
	intersection angles α_{ij}	**just right**	*existence not guaranteed for all triangulations*
METRIC (Section 5)	edge lengths ℓ_{ij}	**just right**	**existence is guaranteed**

Acknowledgments

Thanks to Feng Luo and Boris Springborn for answering questions about discrete uniformization, and to the anonymous referees for providing additional useful comments and references.

References

[AMY18] Paul M. Alsing, Warner A. Miller, and Shing-Tung Yau, *A realization of Thurston's geometrization: discrete Ricci flow with surgery*, Ann. Math. Sci. Appl. **3** (2018), no. 1, 31–45, DOI 10.4310/AMSA.2018.v3.n1.a2. MR3781260

[BCGB08] Mirela Ben-Chen, Craig Gotsman, and Guy Bunin, *Conformal Flattening by Curvature Prescription and Metric Scaling*, CG Forum (2008).

[BCK18] Alex Baden, Keenan Crane, and Misha Kazhdan, *Möbius Registration*, Computer Graphics Forum (SGP) **37** (2018), no. 5.

[BDS17] Alexander I. Bobenko, Nikolay Dimitrov, and Stefan Sechelmann, *Discrete uniformization of polyhedral surfaces with non-positive curvature and branched covers over the sphere via hyper-ideal circle patterns*, Discrete Comput. Geom. **57** (2017), no. 2, 431–469, DOI 10.1007/s00454-016-9830-2. MR3602861

[BG17] Alexander I. Bobenko and Felix Günther, *Discrete Riemann surfaces based on quadrilateral cellular decompositions*, Adv. Math. **311** (2017), 885–932, DOI 10.1016/j.aim.2017.03.010. MR3628233

[BHS06] Alexander I. Bobenko, Tim Hoffmann, and Boris A. Springborn, *Minimal surfaces from circle patterns: geometry from combinatorics*, Ann. of Math. (2) **164** (2006), no. 1, 231–264, DOI 10.4007/annals.2006.164.231. MR2233848

[BPS15] Alexander I. Bobenko, Ulrich Pinkall, and Boris A. Springborn, *Discrete conformal maps and ideal hyperbolic polyhedra*, Geom. Topol. **19** (2015), no. 4, 2155–2215, DOI 10.2140/gt.2015.19.2155. MR3375525

[Bra92] Walter Brägger, *Kreispackungen und Triangulierungen* (German), Enseign. Math. (2) **38** (1992), no. 3-4, 201–217. MR1189006

[BS04a] Alexander I. Bobenko and Boris A. Springborn, *Variational principles for circle patterns and Koebe's theorem*, Trans. Amer. Math. Soc. **356** (2004), no. 2, 659–689, DOI 10.1090/S0002-9947-03-03239-2. MR2022715

[BS04b] Philip L. Bowers and Kenneth Stephenson, *Uniformizing dessins and Belyĭ maps via circle packing*, Mem. Amer. Math. Soc. **170** (2004), no. 805, xii+97, DOI 10.1090/memo/0805. MR2053391

[BS07] Alexander I. Bobenko and Boris A. Springborn, *A discrete Laplace-Beltrami operator for simplicial surfaces*, Discrete Comput. Geom. **38** (2007), no. 4, 740–756, DOI 10.1007/s00454-007-9006-1. MR2365833

[Büc16] Ulrike Bücking, *Approximation of conformal mappings using conformally equivalent triangular lattices*, Advances in discrete differential geometry, Springer, [Berlin], 2016, pp. 133–149. MR3587184

[Büc18a] Ulrike Bücking, *C^∞-convergence of conformal mappings for conformally equivalent triangular lattices*, Results Math. **73** (2018), no. 2, Art. 84, 21, DOI 10.1007/s00025-018-0845-2. MR3806073

[Büc18b] ———, *Conformally symmetric triangular lattices and discrete ϑ-conformal maps*, arXiv e-prints (2018), arXiv:1808.08064.

[BV04] Stephen Boyd and Lieven Vandenberghe, *Convex optimization*, Cambridge University Press, Cambridge, 2004. MR2061575

[CdGDS13] Keenan Crane, Fernando de Goes, Mathieu Desbrun, and Peter Schröder, *Digital geometry processing with discrete exterior calculus*, ACM SIGGRAPH 2013 Courses, SIGGRAPH '13, ACM, 2013.

[Cer96] Davide P. Cervone, *Tight immersions of simplicial surfaces in three space*, Topology **35** (1996), no. 4, 863–873, DOI 10.1016/0040-9383(95)00047-X. MR1404913

[CL03] Bennett Chow and Feng Luo, *Combinatorial Ricci flows on surfaces*, J. Differential Geom. **63** (2003), no. 1, 97–129. MR2015261

[Col91] Yves Colin de Verdière, *Un principe variationnel pour les empilements de cercles* (French), Invent. Math. **104** (1991), no. 3, 655–669, DOI 10.1007/BF01245096. MR1106755

[Cra13] Keenan Crane, *Conformal geometry processing*, Ph.D. thesis, Caltech, June 2013.

[CS03] Charles R. Collins and Kenneth Stephenson, *A circle packing algorithm*, Comput. Geom. **25** (2003), no. 3, 233–256, DOI 10.1016/S0925-7721(02)00099-8. MR1975216

[CSD02] David Cohen-Steiner and Mathieu Desbrun, *Hindsight: LSCM and DNCP are One and the Same*, 2002.

[CW17] Keenan Crane and Max Wardetzky, *A glimpse into discrete differential geometry*, Notices Amer. Math. Soc. **64** (2017), no. 10, 1153–1159, DOI 10.1090/noti1578. MR3701147

[CZ17] M. Campen and D. Zorin, *On Discrete Conformal Seamless Similarity Maps*, ArXiv e-prints (2017).

[DGL08] Junfei Dai, Xianfeng David Gu, and Feng Luo, *Variational principles for discrete surfaces*, Advanced Lectures in Mathematics (ALM), vol. 4, International Press, Somerville, MA; Higher Education Press, Beijing, 2008. MR2439807

[DHLM05] M. Desbrun, A. N. Hirani, M. Leok, and J. E. Marsden, *Discrete Exterior Calculus*, ArXiv Mathematics e-prints (2005).

[DMA02] Mathieu Desbrun, Mark Meyer, and Pierre Alliez, *Intrinsic parameterizations of surface meshes*, Computer Graphics Forum **21** (2002), no. 3, 209–218.

[Fil08] François Fillastre, *Polyhedral hyperbolic metrics on surfaces*, Geom. Dedicata **134** (2008), 177–196, DOI 10.1007/s10711-008-9252-2. MR2399657

[Ge18] Huabin Ge, *Combinatorial Calabi flows on surfaces*, Trans. Amer. Math. Soc. **370** (2018), no. 2, 1377–1391, DOI 10.1090/tran/7196. MR3729504

[Gen02] Adrian P. Gentle, *Regge calculus: a unique tool for numerical relativity*, Gen. Relativity Gravitation **34** (2002), no. 10, 1701–1718, DOI 10.1023/A:1020128425143. MR1938626

[GGL+18] Xianfeng Gu, Ren Guo, Feng Luo, Jian Sun, and Tianqi Wu, *A discrete uniformization theorem for polyhedral surfaces II*, J. Differential Geom. **109** (2018), no. 3, 431–466, DOI 10.4310/jdg/1531188190. MR3825607

[Gli11] David Glickenstein, *Discrete conformal variations and scalar curvature on piecewise flat two- and three-dimensional manifolds*, J. Differential Geom. **87** (2011), no. 2, 201–237. MR2788656

[GLSW18] Xianfeng David Gu, Feng Luo, Jian Sun, and Tianqi Wu, *A discrete uniformization theorem for polyhedral surfaces*, J. Differential Geom. **109** (2018), no. 2, 223–256, DOI 10.4310/jdg/1527040872. MR3807319

[GMMD14] Fernando de Goes, Pooran Memari, Patrick Mullen, and Mathieu Desbrun, *Weighted triangulations for geometry processing*, ACM Trans. Graph. **33** (2014), no. 3, 28:1–28:13.

[Guo11] Ren Guo, *Local rigidity of inversive distance circle packing*, Trans. Amer. Math. Soc. **363** (2011), no. 9, 4757–4776, DOI 10.1090/S0002-9947-2011-05239-6. MR2806690

[GX15] Huabin Ge and Xu Xu, *A discrete Ricci flow on surfaces with hyperbolic background geometry*, Int. Math. Res. Not. IMRN **11** (2017), 3510–3527. MR3693657

[GY08] Xianfeng David Gu and Shing-Tung Yau, *Computational conformal geometry*, Advanced Lectures in Mathematics (ALM), vol. 3, International Press, Somerville, MA; Higher Education Press, Beijing, 2008. With 1 CD-ROM (Windows, Macintosh and Linux). MR2439718

[Ham82] Richard S. Hamilton, *Three-manifolds with positive Ricci curvature*, J. Differential Geometry **17** (1982), no. 2, 255–306. MR664497

[Hat02] Allen Hatcher, *Algebraic topology*, Cambridge University Press, Cambridge, 2002. MR1867354

[HK15] J. Hass and P. Koehl, *A Metric for genus-zero surfaces*, arXiv e-prints (2015).

[HR93] Craig D. Hodgson and Igor Rivin, *A characterization of compact convex polyhedra in hyperbolic* 3*-space*, Invent. Math. **111** (1993), no. 1, 77–111, DOI 10.1007/BF01231281. MR1193599

[HS15] Joel Hass and Peter Scott, *Simplicial energy and simplicial harmonic maps*, Asian J. Math. **19** (2015), no. 4, 593–636, DOI 10.4310/AJM.2015.v19.n4.a2. MR3423736

[Hut91] John E. Hutchinson, *Computing conformal maps and minimal surfaces*, Workshop on Theoretical and Numerical Aspects of Geometric Variational Problems (Canberra, 1990), Proc. Centre Math. Appl. Austral. Nat. Univ., vol. 26, Austral. Nat. Univ., Canberra, 1991, pp. 140–161. MR1139035

[IBI08] Alexander I. Bobenko and Ivan Izmestiev, *Alexandrov's theorem, weighted Delaunay triangulations, and mixed volumes* (English, with English and French summaries), Ann. Inst. Fourier (Grenoble) **58** (2008), no. 2, 447–505. MR2410380

[ILTC01] C. Indermitte, Th. M. Liebling, M. Troyanov, and H. Clémençon, *Voronoi diagrams on piecewise flat surfaces and an application to biological growth*, Theoret. Comput. Sci. **263** (2001), no. 1-2, 263–274, DOI 10.1016/S0304-3975(00)00248-6. Combinatorics and computer science (Palaiseau, 1997). MR1846934

[JKG07] Miao Jin, Junho Kim, and Xianfeng David Gu, *Discrete surface ricci flow: Theory and applications*, Mathematics of Surfaces XII (Berlin, Heidelberg) (Ralph Martin, Malcolm Sabin, and Joab Winkler, eds.), Springer Berlin Heidelberg, 2007, pp. 209–232.

[KCD+16] Mina Konakovic, Keenan Crane, Bailin Deng, Sofien Bouaziz, Daniel Piker, and Mark Pauly, *Beyond developable: Computational design and fabrication with auxetic materials*, ACM Trans. Graph. **35** (2016), no. 4.

[KLRR18] Richard Kenyon, Wai Yeung Lam, Sanjay Ramassamy, and Marianna Russkikh, *Dimers and Circle patterns*, arXiv e-prints (2018), arXiv:1810.05616.

[KPD09] Daniel Kane, Gregory N. Price, and Erik D. Demaine, *A pseudopolynomial algorithm for alexandrov's theorem*, Algorithms and Data Structures (Berlin, Heidelberg) (Frank Dehne, Marina Gavrilova, Jörg-Rüdiger Sack, and Csaba D. Tóth, eds.), Springer Berlin Heidelberg, 2009, pp. 435–446.

[KSS06] Liliya Kharevych, Boris Springborn, and Peter Schröder, *Discrete conformal mappings via circle patterns*, ACM Trans. Graph. **25** (2006), no. 2, 412–438.

[Lis17] Marcin Lis, *Circle patterns and critical Ising models*, Comm. Math. Phys. **370** (2019), no. 2, 507–530, DOI 10.1007/s00220-019-03541-1. MR3994578

[LP16] Wai Yeung Lam and Ulrich Pinkall, *Holomorphic vector fields and quadratic differentials on planar triangular meshes*, Advances in discrete differential geometry, Springer, [Berlin], 2016, pp. 241–265. MR3587188

[LP18] Wai Yeung Lam and Ulrich Pinkall, *Infinitesimal conformal deformations of triangulated surfaces in space*, Discrete Comput. Geom. **60** (2018), no. 4, 831–858, DOI 10.1007/s00454-018-0008-y. MR3869452

[LPRM02] Bruno Lévy, Sylvain Petitjean, Nicolas Ray, and Jérome Maillot, *Least Squares Conformal Maps*, ACM Transactions on Graphics **21** (2002), no. 3.

[Luo04] Feng Luo, *Combinatorial Yamabe flow on surfaces*, Commun. Contemp. Math. **6** (2004), no. 5, 765–780, DOI 10.1142/S0219199704001501. MR2100762

[Luo11] Feng Luo, *Rigidity of polyhedral surfaces, III*, Geom. Topol. **15** (2011), no. 4, 2299–2319, DOI 10.2140/gt.2011.15.2299. MR2862158

[Mer01] Christian Mercat, *Discrete Riemann surfaces and the Ising model*, Comm. Math. Phys. **218** (2001), no. 1, 177–216, DOI 10.1007/s002200000348. MR1824204

[MGA$^+$17] Haggai Maron, Meirav Galun, Noam Aigerman, Miri Trope, Nadav Dym, Ersin Yumer, Vladimir G. Kim, and Yaron Lipman, *Convolutional neural networks on surfaces via seamless toric covers*, ACM Trans. Graph. **36** (2017), no. 4.

[Mil94] John Milnor, *Collected papers. Vol. 1*, Publish or Perish, Inc., Houston, TX, 1994. Geometry. MR1277810

[MTAD08] Patrick Mullen, Yiying Tong, Pierre Alliez, and Mathieu Desbrun, *Spectral conformal parameterization*, Proceedings of the Symposium on Geometry Processing (Aire-la-Ville, Switzerland, Switzerland), SGP '08, Eurographics Association, 2008, pp. 1487–1494.

[Pen87] R. C. Penner, *The decorated Teichmüller space of punctured surfaces*, Comm. Math. Phys. **113** (1987), no. 2, 299–339. MR919235

[PP93] Ulrich Pinkall and Konrad Polthier, *Computing discrete minimal surfaces and their conjugates*, Experiment. Math. **2** (1993), no. 1, 15–36. MR1246481

[Reg61] T. Regge, *General relativity without coordinates* (English, with Italian summary), Nuovo Cimento (10) **19** (1961), 558–571. MR127372

[Riv93] Igor Rivin, *On geometry of convex ideal polyhedra in hyperbolic 3-space*, Topology **32** (1993), no. 1, 87–92, DOI 10.1016/0040-9383(93)90039-X. MR1204408

[Riv94] Igor Rivin, *Euclidean structures on simplicial surfaces and hyperbolic volume*, Ann. of Math. (2) **139** (1994), no. 3, 553–580, DOI 10.2307/2118572. MR1283870

[Riv96] Igor Rivin, *A characterization of ideal polyhedra in hyperbolic 3-space*, Ann. of Math. (2) **143** (1996), no. 1, 51–70, DOI 10.2307/2118652. MR1370757

[Roh11] Steffen Rohde, *Oded Schramm: from circle packing to SLE*, Ann. Probab. **39** (2011), no. 5, 1621–1667, DOI 10.1214/10-AOP590. MR2884870

[RS87] Burt Rodin and Dennis Sullivan, *The convergence of circle packings to the Riemann mapping*, J. Differential Geom. **26** (1987), no. 2, 349–360. MR906396

[RW84] M. Roček and R. M. Williams, *The quantization of Regge calculus*, Z. Phys. C **21** (1984), no. 4, 371–381, DOI 10.1007/BF01581603. MR734698

[SC17] Rohan Sawhney and Keenan Crane, *Boundary First Flattening*, ACM Trans. Graph. **36** (2017), no. 5.

[SC18] Nick Sharp and Keenan Crane, *Variational surface cutting*, ACM Trans. Graph. **37** (2018), no. 4.

[Sch08] Jean-Marc Schlenker, *Circle patterns on singular surfaces* (English, with English and French summaries), Discrete Comput. Geom. **40** (2008), no. 1, 47–102, DOI 10.1007/s00454-007-9045-7. MR2429649

[SLMB05] Alla Sheffer, Bruno Lévy, Maxim Mogilnitsky, and Alexander Bogomyakov, *ABF++: Fast and Robust Angle Based Flattening*, ACM Trans. Graph. **24** (2005), no. 2.

[Smi10] Stanislav Smirnov, *Discrete complex analysis and probability*, Proceedings of the International Congress of Mathematicians. Volume I, Hindustan Book Agency, New Delhi, 2010, pp. 595–621. MR2827906

[Spr08] Boris A. Springborn, *A variational principle for weighted Delaunay triangulations and hyperideal polyhedra*, J. Differential Geom. **78** (2008), no. 2, 333–367. MR2394026

[Spr17] B. Springborn, *Hyperbolic polyhedra and discrete uniformization*, ArXiv e-prints (2017).

[SS00] A. Sheffer and E. De Sturler, *Surface parameterization for meshing by triangulation flattening*, Proc. 9th International Meshing Roundtable, 2000, pp. 161–172.

[SSP08] Boris Springborn, Peter Schröder, and Ulrich Pinkall, *Conformal Equivalence of Triangle Meshes*, ACM Trans. Graph. **27** (2008), no. 3.

[Ste03] Kenneth Stephenson, *Circle packing: a mathematical tale*, Notices Amer. Math. Soc. **50** (2003), no. 11, 1376–1388. MR2011604

[Ste05] Kenneth Stephenson, *Introduction to circle packing*, Cambridge University Press, Cambridge, 2005. The theory of discrete analytic functions. MR2131318

[SWGL15] Jian Sun, Tianqi Wu, Xianfeng Gu, and Feng Luo, *Discrete conformal deformation: algorithm and experiments*, SIAM J. Imaging Sci. **8** (2015), no. 3, 1421–1456, DOI 10.1137/141001986. MR3369062

[Thu79] William Thurston, *Geometry and Topology of 3-Manifolds*, Princeton University, 1979.

[Thu98] William P. Thurston, *Shapes of polyhedra and triangulations of the sphere*, The Epstein birthday schrift, Geom. Topol. Monogr., vol. 1, Geom. Topol. Publ., Coventry, 1998, pp. 511–549, DOI 10.2140/gtm.1998.1.511. MR1668340

[Tro91] Marc Troyanov, *Prescribing curvature on compact surfaces with conical singularities*, Trans. Amer. Math. Soc. **324** (1991), no. 2, 793–821, DOI 10.2307/2001742. MR1005085

[VMW15] Amir Vaxman, Christian Müller, and Ofir Weber, *Conformal mesh deformations with möbius transformations*, ACM Trans. Graph. **34** (2015), no. 4, 55:1–55:11.

[Wil05] S. O. Wilson, *Geometric Structures on the Cochains of a Manifold*, ArXiv Mathematics e-prints (2005).

[WMKG07] Max Wardetzky, Saurabh Mathur, Felix Kaelberer, and Eitan Grinspun, *Discrete Laplace Operators: No Free Lunch*, Proceedings of the Symposium on Geometry Processing (Alexander Belyaev and Michael Garland, eds.), The Eurographics Association, 2007.

[WP11] Johannes Wallner and Helmut Pottmann, *Geometric computing for freeform architecture*, J. Math. Ind. **1** (2011), Art. 4, 18, DOI 10.1186/2190-5983-1-4. MR2824230

[Wu14] Tianqi Wu, *Finiteness of switches in discrete yamabe flow*, Master's thesis, Tsinghua University, May 2014.

[Xu18] Xu Xu, *Rigidity of inversive distance circle packings revisited*, Adv. Math. **332** (2018), 476–509, DOI 10.1016/j.aim.2018.05.026. MR3810260

[ZGLG12] Wei Zeng, Ren Guo, Feng Luo, and Xianfeng Gu, *Discrete heat kernel determines discrete riemannian metric*, Graph. Models **74** (2012), no. 4, 121–129.

[ZGZ+14] M. Zhang, R. Guo, W. Zeng, F. Luo, S.-T. Yau, and X. Gu, *The Unified Surface Ricci Flow*, ArXiv e-prints (2014).

[ZLH+18] Hui Zhao, Xuan Li, Huabin Ge, Na Lei, Min Zhang, Xiaoling Wang, and Xianfeng Gu, *Conformal mesh parameterization using discrete Calabi flow*, Comput. Aided Geom. Design **63** (2018), 96–108, DOI 10.1016/j.cagd.2018.03.001. MR3808380

[ZLS07] Rhaleb Zayer, Bruno Lévy, and Hans-Peter Seidel, *Linear Angle Based Parameterization*, Proc. Symp. Geom. Proc., 2007, pp. 135–141.

Carnegie Mellon University, 5000 Forbes Ave, Pittsburgh, Pennsylvania 15213
Email address: kmcrane@cs.cmu.edu

Proceedings of Symposia in Applied Mathematics
Volume **76**, 2020
https://doi.org/10.1090/psapm/076/00673

Optimal transport on discrete domains

Justin Solomon

Abstract. Inspired by the matching of supply to demand in logistical problems, the optimal transport (or Monge–Kantorovich) problem involves the matching of probability distributions defined over a geometric domain like a surface or manifold. In its most obvious discretization, optimal transport becomes a large-scale linear program, which typically is infeasible to solve efficiently on triangle meshes, graphs, point clouds, and other domains encountered in graphics and machine learning. Recent breakthroughs in numerical optimal transport, however, enable scalability to orders-of-magnitude larger problems, solvable in a fraction of a second. Here, we discuss advances in numerical optimal transport that leverage understanding of both discrete and smooth aspects of the problem. State-of-the-art techniques in discrete optimal transport combine insight from partial differential equations (PDE) with convex analysis to reformulate, discretize, and optimize transportation problems. The end result is a set of theoretically-justified models suitable for domains with thousands or millions of vertices. Since numerical optimal transport is a relatively new discipline, special emphasis is placed on identifying and explaining open problems in need of mathematical insight and additional research.

1. Introduction

Many tools from discrete differential geometry (DDG) were inspired by practical considerations in areas like computer graphics and vision. Disciplines like these require fine-grained understanding of geometric structure and the relationships between different shapes—problems for which the toolbox from smooth geometry can provide substantial insight. Indeed, a triumph of discrete differential geometry is its incorporation into a wide array of computational pipelines, affecting the way artists, engineers, and scientists approach problem-solving across geometry-adjacent disciplines.

A key but neglected consideration hampering adoption of ideas in DDG in fields like computer vision and machine learning, however, is *resilience* to noise and uncertainty. The view of the world provided by video cameras, depth sensors, and other equipment is extremely unreliable. Shapes do not necessarily come to a

2010 *Mathematics Subject Classification.* Primary 90C99; Secondary 90B06.

Key words and phrases. Discrete differential geometry, optimal transport.

The author was supported by Army Research Office grant W911NF-12-R-0011 ("Smooth Modeling of Flows on Graphs"), by National Science Foundation grant IIS-1838071 ("BIG-DATA:F: Statistical and Computational Optimal Transport for Geometric Data Analysis"), by the MIT Research Support Committee ("Structured Optimization for Geometric Problems"), and by the MIT–IBM Watson AI Lab ("Large-Scale Optimal Transport for Machine Learning").

computer as complete, manifold meshes but rather may be scattered clouds of points that represent, e.g., only those features visible from a single position. Similarly, it may be impossible to pinpoint a feature on a shape exactly; rather, we may receive only a fuzzy signal indicating where a point or feature of interest *may* be located. Such uncertainty only increases in high-dimensional statistical contexts, where the presence of geometric structure in a given dataset is itself not a given. Rather than regarding this messiness as an "implementation issue" to be coped with by engineers adapting DDG to imperfect data, however, the challenge of developing principled yet noise-resilient discrete theories of shape motivates new frontiers in mathematical research.

Probabilistic language provides a natural means of formalizing notions of uncertainty in the geometry processing pipeline. In place of representing a feature or shape directly, we might instead use a probability distribution to encode a rougher notion of shape. Unfortunately, this proposal throws both smooth and discrete constructions off their foundations: We must return to the basics and redefine notions like distance, distortion, and curvature in a fashion that does not rely on knowing shape with infinite precision and confidence. At the same time, we must prove that the classical case is recovered as uncertainty diminishes to zero.

The mathematical discipline of *optimal transport* (OT) shows promise for making geometry work in the probabilistic regime. In its most basic form, optimal transport provides a means of lifting distances between points on a domain to distances between probability distributions *over* the domain. The basic construction of OT is to interpret probability distributions as piles of sand; the distance between two such piles of sand is defined as the amount of work it takes to transform one pile into the other. This intuitive construction gave rise to an alternative name for OT in the computational world: The "earth mover's distance" (EMD) [**RTG00**]. Indeed, the basic approach in OT is so natural that it has been proposed and re-proposed in many forms and with many names, from OT to EMD, the Mallows distance [**LB01**], the Monge–Kantorovich problem [**Vil03**], the Hitchcock–Koopmans transportation problem [**Hit41**, **Koo41**], the Wasserstein/Vaseršteĭn distance [**Vas69**, **Dob70**], and undoubtedly many others.

Many credit Gaspard Monge with first formalizing the optimal transport problem in 1781 [**Mon81**]. Beyond its early history, modern understanding of optimal transport dates back only to the World War II era, through the Nobel Prize-winning work of Leonid Kantorovich [**Kan42**]. Jumping forward several decades, while many branches of DDG are dedicated to making centuries-old constructions on smooth manifolds work in the discrete case, optimal transport has the distinction of continuing to be an active area of research in the mathematical community whose basic properties are still being discovered. Indeed, the computational and theoretical literature in this area move in lock-step: New theoretical constructions often are adapted by the computational community in a matter of months, and some key theoretical ideas in transport were inspired by computational considerations and constructions.

Here, we aim to provide some intuition about transport and its relevance to the discrete differential geometry world. While a complete survey of work on OT or even just its computational aspects is worthy of a full textbook, here we focus on the narrower problem of how to "make transport work" on a discretized piece of geometry amenable to representation on a computer. The primary aim is to highlight the challenges in transitioning from smooth to discrete, to illustrate some basic constructions that have been proposed recently for this task, and—most importantly—to expose the plethora of open questions remaining in the relatively young discipline of computational OT. No-doubt incomplete references are provided to selected intriguing ideas in computational OT, each of which is worthy of far more detailed discussion.

Additional reference. Those readers with limited experience in related disciplines may wish to begin by reading [**Sol17**], a shorter survey by the author on the same topic, intended for a generalist audience.

Disclaimer. These notes are intended as a short, intuitive, and *extremely* informal introduction. Optimal transport is a popular topic in mathematical research, and interested readers should refer to surveys such as [**Vil03, Vil08**] for more comprehensive discussion. The recent text [**San15**] provides discussion targeted to the applied world. A few recent surveys also are targeted to computational issues in optimal transport [**LS18, PC19**].

The author of this tutorial offers his sincere apology to those colleagues whose foundational work is undoubtedly yet accidentally omitted from this document. A "venti"-sized caffeinated beverage is humbly offered in exchange for those readers' forgiveness and understanding.

2. Motivation: From probability to discrete geometry

To motivate the construction of optimal transport in the context of geometry processing, we begin by considering the case of smooth probability distributions over the real numbers $\mathbb{R}$. Here, the geometry is extremely simple, described by values $x \in \mathbb{R}$ equipped with the distance metric $d(x, y) := |x - y|$. Then we expand to define the transport problem in more generality and state a few useful properties.

2.1. The transport problem. Define the space of probability measures over $\mathbb{R}$ as $\mathrm{Prob}(\mathbb{R})$. Without delving into the formalities of measure theory (see [**Jon01, Coh80**] for formal introductions), these are roughly the functions $\mu \in \mathrm{Prob}(\mathbb{R})$ assigning probabilities to subsets $S \subseteq \mathbb{R}$ such that $\mu(S) \geq 0$ for all measurable S, $\mu(\mathbb{R}) = 1$, and $\mu(\cup_{i=1}^k S_i) = \sum_{i=1}^k \mu(S_i)$ for disjoint sets $\{S_i \subseteq \mathbb{R}\}_{i=1}^k$. Under the stronger assumption of *absolute continuity*, μ admits a *distribution function (or probability density function)* $\rho(x) : \mathbb{R} \to \mathbb{R}$ assigning a probability density to every point:

$$\mu(S) = \int_S \rho(x)\, dx.$$

The function $\rho(x)$ reflects the familiar notion of a *probability distribution*, like a Gaussian bell curve over the real numbers $\mathbb{R}$.

Measure theory, probability, and statistics each are constructed from slightly different interpretations of the set of probability distributions $\mathrm{Prob}(\mathbb{R})$. Adding to the mix, we can think of optimal transport as a *geometric* theory of probability. In particular, as illustrated in Figure 1, roughly a probability distribution over $\mathbb{R}$ can be thought of as a superposition of points in $\mathbb{R}$, weighted by the function ρ. We can recover a (complicated) representation for a single point $x \in \mathbb{R}$ as a *Dirac δ-measure* centered at x, a probability distribution whose complete mass is concentrated at x.

Applying a physical perspective, we can also think of distributions geometrically using a physical analogy. Suppose we are given a bucket of sand whose total mass

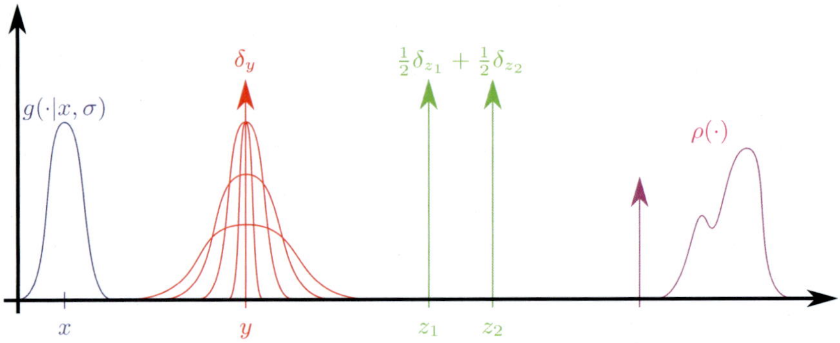

FIGURE 1. One-dimensional examples of probability distributions used to encode geometric features with uncertainty. A probability distribution like a Gaussian g with mean x and standard deviation σ can be thought of as a "fuzzy" location of a point in $x \in \mathbb{R}$ with uncertainty σ. As a distribution sharpens about its mean to a δ-function δ_y, it encodes a classical piece of geometry: a point $y \in \mathbb{R}$. Distributional language, however, is fundamentally broader, including constructions like the superposition of two points z_1 and z_2 or combining a point and a fuzzy feature into a single distribution ρ.

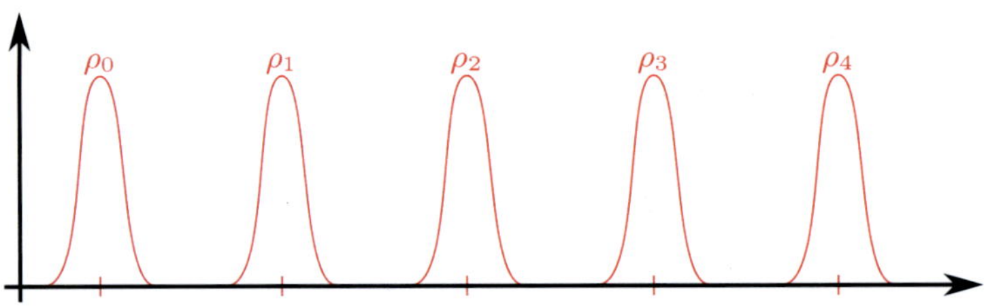

FIGURE 2. The distributions $\rho_0, \ldots, \rho_4$ are equidistant with respect to the L_1 and KL divergences, while the Wasserstein distance from optimal transport increases linearly with distance over $\mathbb{R}$.

is one pound. We could distribute this pound of sand across the real numbers by stacking it all at a single point (a δ-measure), concentrating it at a few points, or spreading it out smoothly. The height of the pile of sand expresses a geometric feature: Lots of sand at a point $x \in \mathbb{R}$ indicates we think a feature is located at x.

If we wish to deepen this analogy and lift notions from geometry to the space $\text{Prob}(\mathbb{R})$, perhaps the most basic object we must define is a notion of *distance* between two distributions $\mu_0, \mu_1 \in \text{Prob}(\mathbb{R})$ that resembles the distance $d(x, y) = |x - y|$ between points on the underlying space. Supposing for now that μ_0 and μ_1 admit density functions ρ_0 and ρ_1, respectively, a few candidate notions of distance

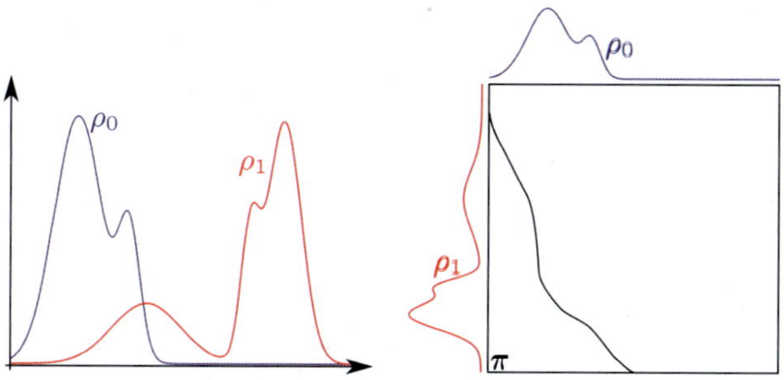

(a) Source and target (b) Transport map

FIGURE 3. Two distributions over the real line (a) and the resulting transport map (b). In (b), the box is the product space $[0,1]\times[0,1]$, and dark values indicate a matching between ρ_0 and ρ_1.

or divergence come to mind:

$$L_1 \text{ distance:} d_{L_1}(\rho_0,\rho_1) := \int_{-\infty}^{\infty} |\rho_0(x)-\rho_1(x)|\,dx$$

$$\text{KL divergence:} d_{\mathrm{KL}}(\rho_0\|\rho_1) := \int_{-\infty}^{\infty} \rho_0(x)\log\frac{\rho_0(x)}{\rho_1(x)}\,dx.$$

These divergences are used widely in analysis and information theory, but they are insufficient for geometric computation. In particular, consider the distributions in Figure 2. The two divergences above give the same value for any pair of different ρ_i's! This is because they measure only the overlap; the ground distance $d(x,y)=|x-y|$ is never used in their computation.

Optimal transport resolves this issue by leveraging the physical analogy proposed above. In particular, suppose our sand is currently in arrangement ρ_0 and we wish to *reshape* it to a new distribution ρ_1. We take a steam shovel and begin scooping up the sand at points x in ρ_0 where $\rho_0(x)>\rho_1(x)$ and dropping it places where $\rho_1(x)>\rho_0(x)$; eventually one distribution is transformed into the other.

There are many ways the steam shovel could approach its task: We could move sand efficiently, or we could drive it miles away and then drive back, wasting fuel in the process. But assuming $\rho_0\neq\rho_1$, there is some amount of work inherent in the fact that ρ_0 and ρ_1 are not the same. We can formalize this idea by solving for an unknown measure $\pi(x,y)$ determining how much mass gets moved from x to y by the steam shovel for each (x,y) pair. The minimum amount of work is then

$$(1)\quad \mathcal{W}_1(\rho_0,\rho_1) := \left\{\begin{array}{ll} \min_\pi \iint_{\mathbb{R}\times\mathbb{R}} \pi(x,y)|x-y|\,dx\,dy & \text{Minimize total work} \\ \quad\text{s.t. } \pi\geq 0\ \forall x,y\in\mathbb{R} & \text{Nonnegative mass} \\ \qquad \int_{\mathbb{R}} \pi(x,y)\,dy = \rho_0(x)\ \forall x\in\mathbb{R} & \text{Starts from } \rho_0 \\ \qquad \int_{\mathbb{R}} \pi(x,y)\,dx = \rho_1(y)\ \forall y\in\mathbb{R} & \text{Ends at } \rho_1. \end{array}\right.$$

This optimization problem quantifies the minimum amount of work—measured as mass $\pi(x,y)$ times distance traveled $|x-y|$—required to transform ρ_0 into ρ_1. We can think of the unknown function π as the instructions given to the laziest possible steam shovel tasked with dropping one distribution onto another. This amount of work is known as the *1-Wasserstein distance* in optimal transport; in one dimension, it equals the L_1 distance between the cumulative distribution functions of ρ_0 and ρ_1. An example of ρ_0, ρ_1, and the resulting π is shown in Figure 3.

Generalizing slightly, we can define the p-Wasserstein distance:

$$(2) \qquad [\mathcal{W}_p(\rho_0,\rho_1)]^p := \left\{ \begin{array}{rl} \min_\pi & \iint_{\mathbb{R}\times\mathbb{R}} \pi(x,y)|x-y|^p\,dx\,dy \\ \text{s.t.} & \pi \geq 0\ \forall x,y\in\mathbb{R} \\ & \int_{\mathbb{R}} \pi(x,y)\,dy = \rho_0(x)\,\forall x\in\mathbb{R} \\ & \int_{\mathbb{R}} \pi(x,y)\,dx = \rho_1(y)\,\forall y\in\mathbb{R}. \end{array} \right.$$

In analogy to Euclidean space, many properties of $\mathcal{W}_p$ are split into cases $p<1$, $p=1$, and $p>1$; for instance, it satisfies the triangle inequality any time $p\geq 1$. The $p=2$ case is of particular interest in the literature and corresponds to a "least-squares" version of transport that minimizes kinetic energy rather than work (see §2.4). Generalizing (2) even more, if we replace $|x-y|^p$ with a generic cost $c(x,y)$ we recover the *Kantorovich problem* [**Kan42**].

It is important to note an alternative formulation of the transport problem (2), which historically was posed first but does not always admit a solution. Rather than optimizing for a function $\pi(x,y)$ with an unknown for every possible (x,y) pair, one could consider an alternative in which instead the variable is a single function $\phi(x)$ that "pushes forward" ρ_0 onto ρ_1; this corresponds to choosing a single destination $\phi(x)$ for every source point x. In this case, the objective function would look like

$$(3) \qquad \int_{-\infty}^{\infty} |\phi(x)-x|^p \rho_0(x)\,dx,$$

and the constraints would ask that the image of ρ_0 under ϕ is ρ_1, notated $\phi_\sharp\rho_0 = \rho_1$. While this version corresponds to the original version of transport proposed by Monge, sometimes for the transport problem to be solvable it is necessary to split the mass at a single source point to multiple destinations. A triumph of theoretical optimal transport, however, shows that $\pi(x,y)$ is nonzero only on a "graph" $\{(x,\phi(x)) : x\in\mathbb{R}\}$ whenever ρ_0 is absolutely continuous, that is, whenever we match probability density functions rather than more general measures, linking the two problems.

2.2. Discrete problems in one dimension. So far our definitions have not been amenable to numerical computation: Our unknowns are functions $\pi(x,y)$ with *infinite* numbers of variables (one value of π for each (x,y) pair in $\mathbb{R}\times\mathbb{R}$)—certainly more than can be stored on a computer with finite capacity. Continuing to work in one dimension, we suggest some special cases where we can solve the transport problem with a finite number of variables.

Rather than working with distribution functions $\rho(x)$, we will relax to the more general case of transport between measures $\mu_0,\mu_1\in\mathrm{Prob}(\mathbb{R})$. Define the Dirac δ-measure centered at $x\in\mathbb{R}$ via

$$\delta_x(S) := \left\{ \begin{array}{ll} 1 & \text{if } x\in S \\ 0 & \text{otherwise.} \end{array} \right.$$

It is easy to check that $\delta_x(\cdot)$ is a probability measure.

Suppose $\mu_0,\mu_1\in\mathrm{Prob}(\mathbb{R})$ can be written as *superpositions* of δ measures:

$$(4) \qquad \mu_0 := \sum_{i=1}^{k_0} a_{0i}\delta_{x_{0i}} \text{and} \mu_1 := \sum_{i=1}^{k_1} a_{1i}\delta_{x_{1i}},$$

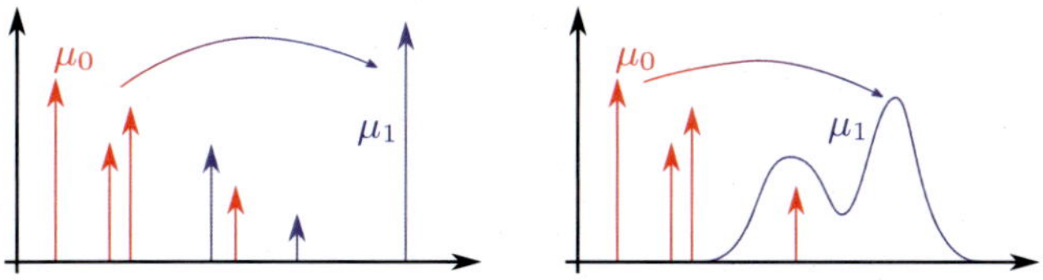

(a) Fully discrete transport (b) Semidiscrete transport

FIGURE 4. Discrete (a) and semidiscrete (b) optimal transport in one dimension.

where $1 = \sum_{i=1}^{k_0} a_{0i} = \sum_{i=1}^{k_1} a_{1i}$ and $a_{0i}, a_{1i} \geq 0$ for all i. Figure 4(a) illustrates this case; all the mass of μ_0 and μ_1 is concentrated at a few isolated points.

In the case where the source and target distributions are composed of δ's, we only can move mass between pairs of points $x_{0i} \mapsto x_{1j}$. Taking T_{ij} the total mass moved from x_{0i} to x_{1j}, we can solve for $\mathcal{W}_p^p$ as

$$[\mathcal{W}_p(\mu_0, \mu_1)]^p = \left\{ \begin{array}{rl} \min_{T \in \mathbb{R}^{k_0 \times k_1}} & \sum_{ij} T_{ij} |x_{0i} - x_{1j}|^p \\ \text{s.t.} & T \geq 0 \\ & \sum_j T_{ij} = a_{0i} \\ & \sum_i T_{ij} = a_{1j}. \end{array} \right. \tag{5}$$

This is an optimization problem in $k_0 k_1$ variables T_{ij}: No need for an infinite number of $\pi(x, y)$'s! In fact, it is a *linear program* solvable using many classic algorithms, such as the simplex or interior point methods [**BV04**].

There is a more subtle case where we can still represent the unknown in optimal transport using a finite number of variables. Suppose $\mu_0 \in \text{Prob}(\mathbb{R})$ is a superposition of δ measures and $\mu_1 \in \text{Prob}(\mathbb{R})$ is absolutely continuous, implying μ_1 admits a distribution function $\rho_1(x)$:

$$\mu_0 := \sum_{i=1}^{k} a_i \delta_{x_i} \text{and} \mu_1(S) := \int_S \rho_1(x)\, dx. \tag{6}$$

This situation is illustrated in Figure 4(b); it corresponds to transporting from a distribution whose mass is concentrated at a few points to a distribution whose distribution is more smooth. In the technical literature, this setup is known as *semidiscrete* transport [**AHA92, GM96, Mér11**].

Returning to the transport problem in (2), in this semidiscrete case we can think of the coupling π as decomposing into a set of measures $\pi_1, \pi_2, \ldots, \pi_k \in \text{Prob}(\mathbb{R})$ where each term in the sum (6) has its own target distribution: $\delta_{x_i} \mapsto \pi_i$. As a sanity check, note that $\mu_1 = \sum_i a_i \pi_i(x)$.

Without loss of generality, we can assume the x_i's are sorted, that is, $x_1 < x_2 < \cdots < x_k$. Suppose $1 \leq i < j \leq k$, and hence $x_i < x_j$. In one dimension, we can convince ourselves that the optimal transport map π should never "leapfrog" mass, that is, the delivery target of the mass at x_i when transported to ρ_1 should be *to the left* of the target of mass at x_j, as illustrated in Figure 5. This monotonicity property implies the existence of intervals $[b_1, c_1], [b_2, c_2], \ldots, [b_k, c_k]$ such that π_i is supported in $[b_i, c_i]$ and $c_i < b_j$ whenever $i < j$; the mass $a_i \delta_{x_i}$ is distributed according to $\rho_1(x)$ in the interval $[b_i, c_i]$.

The semidiscrete transport problem corresponds to another case where we can solve a transport problem with a finite number of variables, the b_i's and c_i's. Of

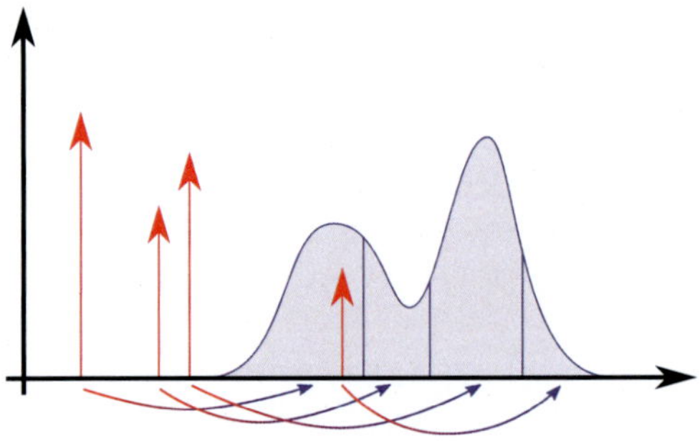

FIGURE 5. Solving 1D semidiscrete transport from Figure 4(b); every Dirac δ-function mass in the source μ_0 gets mapped to a contiguous interval worth of mass in the target μ_1.

course, in one dimension these can be read off from the cumulative distribution function (CDF) of ρ_1, but in higher dimensions this will not be the case. Instead, the intervals $[b_i, c_i]$ will be replaced with *power cells*, a generalization of a Voronoi diagram (§4.3).

While our discussion above gives two cases in which a computer could plausibly solve the transport problem, they do not correspond to the usual situation for DDG in which the geometry itself—in this case the real line $\mathbb{R}$—is discretized. As we will see in future sections, there currently does not exist consensus about what to do in this case but several possible adaptations have been proposed.

2.3. Moving to higher dimensions. We are now ready to state the optimal transport problem in full generality; this transition will require somewhat more complicated notation but builds on the one-dimensional intuition introduced in previous sections. Following [**Vil03**, §1.1.1], take (X, μ) and (Y, ν) to be probability spaces, paired with a nonnegative measurable cost function $c(x, y)$. Define a *measure coupling* $\pi \in \Pi(\mu, \nu)$ as follows:

DEFINITION 1 (Measure coupling). A *measure coupling* $\pi \in \text{Prob}(X \times Y)$ is a probability measure on $X \times Y$ satisfying

$$\pi(A \times Y) = \mu(A)$$
$$\pi(X \times B) = \nu(B)$$

for all measurable $A \subseteq X$ and $B \subseteq Y$. The set of measure couplings between μ and ν is denoted $\Pi(\mu, \nu)$.

With this piece of notation, we can write the Kantorovich optimal transport problem as follows:

$$\boxed{\text{OT}(\mu, \nu; c) := \min_{\pi \in \Pi(\mu,\nu)} \iint_{X \times Y} c(x, y)\, d\pi(x, y)} \tag{7}$$

Here, we use some notation from measure theory: $d\pi(x, y)$ denotes integration against probability measure π. If π admits a distribution function $p(x, y)$, then we can write $d\pi(x, y) = p(x, y)\, dx\, dy$; the more general notation allows for δ measures and other objects that cannot be written as functions.

We highlight a few interesting special cases below:

Discrete transportation. Suppose $X = \{1, 2, \ldots, k_1\}$ and $Y = \{1, 2, \ldots, k_2\}$. Then, $\mu \in \operatorname{Prob}(X)$ can be represented as a vector $v \in \mathbb{S}_{k_1}$ and $\nu \in \operatorname{Prob}(Y)$ can be represented as a vector $w \in \mathbb{S}_{k_2}$, where $\mathbb{S}_k$ denotes the k-dimensional probability simplex:

$$\mathbb{S}_k := \left\{ v \in \mathbb{R}^k : v \geq 0 \text{ and } \sum_i v_i = 1 \right\}. \tag{8}$$

The cost function becomes discrete as well and can be written as a matrix $C = (c_{ij})$. After simplification, the transport problem between $v \in \mathbb{S}_{k_1}$ and $w \in \mathbb{S}_{k_2}$ given cost matrix C becomes

$$\operatorname{OT}(v, w; C) = \begin{cases} \min_{T \in \mathbb{R}^{k_1 \times k_2}} & \sum_{ij} T_{ij} c_{ij} \\ \text{s.t.} & T \geq 0 \\ & \sum_j T_{ij} = v_i \ \forall i \in \{1, \ldots, k_1\} \\ & \sum_i T_{ij} = w_j \ \forall j \in \{1, \ldots, k_2\}. \end{cases} \tag{9}$$

This linear program—which has (5) as a special case—is solvable computationally and is the most obvious way to make optimal transport work in a discrete context. It was proposed in the computational literature as the "earth mover's distance" (EMD) [**RTG00**]. When $k_1 = k_2 := k$ and C is symmetric, nonnegative, and satisfies the triangle inequality, one can check that $\operatorname{OT}(\cdot, \cdot; C)$ is a distance on $\mathbb{S}_k$; see [**CA14**] for a clear proof of this property.

Wasserstein distance. Next, suppose $X = Y = \mathbb{R}^n$, and take $c_{n,p}(x, y) := \|x - y\|_2^p$. Then, we recover the *Wasserstein distance* on $\operatorname{Prob}(\mathbb{R}^n)$, defined via

$$\mathcal{W}_p(\mu, \nu) := [\operatorname{OT}(\mu, \nu; c_{n,p})]^{1/p}. \tag{10}$$

$\mathcal{W}_p$ is a distance when $p \geq 1$, and $\mathcal{W}_p^p$ is a distance when $p \in [0, 1)$ [**Vil03**, §7.1.1]. In fact, the Wasserstein distance can be defined for probability measures over a surface, Riemannian manifold, or even a Polish space via the same formula.

The Wasserstein distance has drawn considerable application-oriented interest and aligns well with the basic motivation laid out in §1. Its basic role is to lift distances between points to distances between distributions in a compatible fashion: The Wasserstein distance between two δ-functions δ_x and δ_y is exactly the distance from x to y. In §3, we will show how this basic property has strong bearing on several computational pipelines that need to lift geometric constructions to uncertain contexts.

2.4. One value, many formulas. A remarkable property of the transport problem (7) is the sheer number of equivalent formulations that all lead to the same value, the cost of transporting mass from one measure onto another. These not only provide many interpretations of the transport problem but also suggest a diverse set of computational algorithms for transport, each of which tackles a different way of writing down the basic problem.

Duality. A basic idea in the world of convex optimization is that of *duality*, that every minimization problem admits a "dual" maximization problem whose optimal value is a lower bound for that of the primal. As with most linear programs, optimal transport typically exhibits *strong duality*: The optimal values of the maximization and minimization problems coincide.

To motivate duality for transport, we will start with the finite-dimensional problem (9). We note two simple identities:

$$\max_{s\in\mathbb{R}} st = \begin{cases} 0 & \text{if } t=0 \\ \infty & \text{otherwise} \end{cases} \quad \max_{s\leq 0} st = \begin{cases} 0 & \text{if } t\geq 0 \\ \infty & \text{otherwise} \end{cases}$$

These allow us to write (9) as follows:

$$\min_T \max_{S\leq 0,\phi,\psi} \left[\sum_{ij} T_{ij}(c_{ij}+S_{ij}) + \sum_i \phi_i \left(v_i - \sum_j T_{ij}\right) + \sum_j \psi_j \left(w_j - \sum_i T_{ij}\right)\right].$$

The dual problem is derived by simply swapping the min and the max:

$$\begin{aligned} &\max_{S\leq 0,\phi,\psi} \min_T \left[\sum_{ij} T_{ij}(c_{ij}+S_{ij}) + \sum_i \phi_i \left(v_i - \sum_j T_{ij}\right) + \sum_j \psi_j \left(w_j - \sum_i T_{ij}\right)\right] \\ &\quad = \max_{S\leq 0,\phi,\psi} \min_T \left[\sum_{ij} T_{ij}(c_{ij}+S_{ij}-\phi_i-\psi_j) + \sum_i \phi_i v_i + \sum_j \psi_j w_j\right] \end{aligned}$$

after refactoring.

Since T is unconstrained in the inner optimization problem of the dual, the solution of the inner minimization is $-\infty$ unless $S_{ij} = \phi_i + \psi_j - c_{ij}$ for all (i,j), that is, unless the coefficient of T_{ij} equals zero. Since the outer problem is a maximization, clearly we should avoid an optimal value of $-\infty$ for the inner minimization. Hence, we can safely add $S_{ij} = \phi_i + \psi_j - c_{ij}$ as a constraint to the dual problem. After some simplification, this extra constraint allows us to arrive at the dual of (9):

$$\begin{array}{rl} \max_{\phi,\psi} & \sum_i [\phi_i v_i + \psi_i w_i] \\ \text{s.t.} & \phi_i + \psi_j \leq c_{ij} \; \forall (i,j). \end{array} \tag{11}$$

Although we have not justified that it is acceptable to swap a max and a min in this context, several techniques ranging from direct proof to the "sledgehammer" Slater duality condition [**Sla50**] show that the optimal value of this maximization problem agrees with the optimal value of the minimization problem (9).

As is often the case in convex optimization, the dual (11) of the transport problem (9) has an intuitive interpretation. Suppose we change roles in optimal transport from the worker who wishes to minimize work to a company that wishes to maximize profit. The customer pays ϕ_i dollars per pound to drop off material v_i to ship from location i and ψ_j dollars per pound to pick up material w_j from location j. The dual problem (11) maximizes profit under the constraint that it is never cheaper for the customer to drive from i to j and ignore the service completely: $\phi_i + \psi_j \leq c_{ij}$.

We pause here to note some rough trade-offs between the primal and dual transport problems. Since both yield the same optimal value, the designer of a computational system for solving optimal transport problems has a decision to make: whether to solve the primal problem, the dual problem, or both simultaneously (the latter aptly named a "primal–dual" algorithm). There are advantages and disadvantages to each approach. The primal problem (9) directly yields the matrix T, which tells not just the cost of transport but how much mass T_{ij} to move from source i to destination j; the only inequality constraint is that the entire matrix has nonnegative entries. On the other hand, the dual problem (11) has fewer variables, making it easier to store the output on the computer, but the "shadow price" variables (ϕ, ψ) are harder to interpret and are constrained by a quadratic number of inequalities. Currently there is little consensus as to which formulation leads to more successful algorithms or discretizations, and state-of-the-art techniques are divided among the two basic approaches.

As with many constructions in optimal transport, the dual of the measure-theoretic problem (7) resembles the discrete case up to a change of the notation. In particular, we can write

$$\text{(12)}\qquad \boxed{\mathrm{OT}(\mu,\nu;c) := \left\{ \begin{array}{rl} \sup_{\substack{\phi\in L^1(d\mu),\\ \psi\in L^1(d\nu)}} & \int_X \phi(x)\,d\mu(x) + \int_Y \psi(y)\,d\nu(y) \\ \text{s.t.} & \phi(x)+\psi(y)\leq c(x,y) \\ & \quad\text{for } d\mu\text{-a.e. } x\in X,\ d\nu\text{-a.e. } y\in Y. \end{array}\right.}$$

This formula comes from a similar min-max swap, justified by results in convex analysis [**Roc15**].

It is worth noting a simplification that appears often in the transport world. Since μ and ν are positive measures and the overall problem in (12) is a maximization, we might as well choose ϕ and ψ as large as possible while satisfying the constraints. Suppose we fix the function $\phi(x)$ and *just* optimize for the function $\psi(x)$. Rearranging the constraint shows that for all $(x, y) \in X \times Y$ we must have $\psi(y) \leq c(x, y) - \phi(x)$. Equivalently, for all $y \in Y$ we must have $\psi(y) \leq \inf_{x\in X}[c(x, y) - \phi(x)]$. Define the *c-transform*

$$\text{(13)}\qquad \phi^c(y) := \inf_{x\in X}[c(x,y) - \phi(x)].$$

By the argument above we have

$$\mathrm{OT}(\mu,\nu;c) = \sup_{\phi\in L^1(d\mu)} \int_X \phi(x)\,d\mu(x) + \int_Y \phi^c(y)\,d\nu(y).$$

This problem is unconstrained, but the transformation from ϕ to ϕ^c is relatively complicated.

We finally simplify one special case of this dual formula, the 1-Wasserstein distance, which has gained recent interest in machine learning thanks to its application in generative adversarial networks (GANs) [**ACB17**]. In this case, $X = Y = \mathbb{R}^n$ and $c(x, y) = \|x - y\|_2$. We can derive a bound as follows:

$$\begin{aligned} |\phi^c(x) - \phi^c(y)| &= \left|\inf_z[\|x-z\|_2 - \phi(z)] - \inf_z[\|y-z\|_2 - \phi(z)]\right| \text{ by definition} \\ &\leq \sup_z \left|\|x-z\|_2 - \|y-z\|_2\right| \\ &\quad \text{by the identity } |\inf_x f(x) - \inf_x g(x)| \leq \sup_x |f(x)-g(x)| \\ &\leq \|x-y\|_2 \text{ by the reverse triangle inequality.} \end{aligned} \tag{14}$$

Furthermore, by definition of the c-transform (13) by taking $x = y$ we have $\phi^c(y) \leq -\phi(y)$, or equivalently $\phi(y) \leq -\phi^c(y)$. Hence,

$$\begin{aligned}
&\mathcal{W}_1(\mu,\nu)\\
&\quad = \mathrm{OT}(\mu,\nu;c) \text{ with the choice } c(x,y) := \|x-y\|_2\\
&\quad = \sup_{\phi\in L^1(d\mu)} \int_{\mathbb{R}^n} \phi(x)\,d\mu(x) + \int_{\mathbb{R}^n} \phi^c(y)\,d\nu(y) \text{ by definition of the } c\text{-transform}\\
&\quad \leq \int_{\mathbb{R}^n} \phi^c(x)\,[d\nu(x) - d\mu(x)] \text{ since } \phi(y) \leq -\phi^c(y)\ \forall y\in\mathbb{R}^n\\
&\quad \leq \sup_{\psi\in\mathrm{Lip}_1(\mathbb{R}^n)} \int_{\mathbb{R}^n} \psi(x)\,[d\nu(x) - d\mu(x)]
\end{aligned}$$

where $\mathrm{Lip}_1(\mathbb{R}^n) := \{f(x) : |f(x) - f(y)| \leq \|x-y\|_2\ \forall x,y\in\mathbb{R}^n\}$.

Lip_1 denotes the set of 1-Lipschitz functions; the last step is derived from (14), which shows that ψ^c is 1-Lipschitz.

In fact, this inequality is an equality. To prove this, take ψ to be any 1-Lipschitz function. Then,

$$\psi^c(y) = \inf_{x\in\mathbb{R}^n} [\|x-y\|_2 - \psi(x)] \geq \inf_{x\in\mathbb{R}^n} [\|x-y\|_2 - \|x-y\|_2 - \psi(y)] = -\psi(y), \tag{15}$$

where we have rearranged the Lipschitz property $\psi(x) - \psi(y) \leq \|x-y\|_2$ to show $-\psi(x) \geq -\|x-y\|_2 - \psi(y)$. Hence,

$$\begin{aligned}
\sup_{\psi\in\mathrm{Lip}_1(\mathbb{R}^n)} \int_{\mathbb{R}^n} \psi(x)\,[d\nu(x) - d\mu(x)] &\leq \sup_{\psi\in\mathrm{Lip}_1(\mathbb{R}^n)} \int_{\mathbb{R}^n} [\psi(x)\,d\nu(x) + \psi^c(y)]\,d\mu(y) \text{ by (15)}\\
&\leq \sup_{\psi\in L^1(d\nu)} \int_{\mathbb{R}^n} [\psi(x)\,d\nu(x) + \psi^c(y)]\,d\mu(y)\\
&\quad\text{since the constraints are loosened}\\
&= \mathcal{W}_1(\mu,\nu).
\end{aligned}$$

This finishes motivating the final formula

$$\boxed{\mathcal{W}_1(\mu,\nu) = \sup_{\psi\in\mathrm{Lip}_1(\mathbb{R}^n)} \int_{\mathbb{R}^n} \psi(x)\,[d\nu(x) - d\mu(x)].}$$

This convenient identity is used in computational contexts because the constraint that a function is 1-Lipschitz is fairly easy to enforce computationally; sadly, it does not extend to other Wasserstein $\mathcal{W}_p$ distances, which have stronger uniqueness and regularity properties when $p > 1$.

Eulerian transport. The language of fluid dynamics introduces two equivalent ways to understand the flow of a liquid or gas as it sloshes in a tank. In the *Lagrangian* framework, the fluid is thought of as a collection of particles whose path we trace as a function of time; the equations of motion roughly determine a map $\Phi_t(x)$ with $\Phi_0(x) = x$ determining the position at time $t \geq 0$ of the particle located at x when $t = 0$. Contrastingly, *Eulerian* fluid dynamics takes the point of view of a barnacle attached to a point in the tank of water counting the number of particles that flow past a point x; this formulation might seek a function $\rho_t(x)$ giving the density of the fluid at a non-moving point x as a function of time t.

So far, our formulation of transport has been Lagrangian: The transportation plan π explicitly determines how to match particles from the source distribution

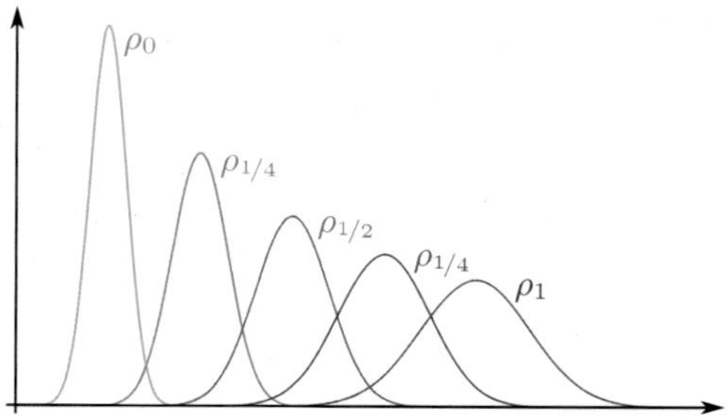

FIGURE 6. Displacement interpolation from ρ_0 to ρ_1 explains optimal transport between these two densities using a time-varying density function ρ_t, $t \in [0, 1]$.

μ to the target distribution ν. Using a particularly clever change of variables, a landmark paper by Benamou & Brenier shows that the 2-Wasserstein distance from (10) over Euclidean space with cost $c(x, y) = \|x - y\|_2^2$ can be computed in an Eulerian fashion [**BB00**]:

$$\mathcal{W}_2^2(\rho_0, \rho_1) = \begin{cases} \min_{v(x,t),\rho(x,t)} \frac{1}{2} \int_0^1 \int_{\mathbb{R}^n} \rho(x,t)\, \|v(x,t)\|_2^2 \, dA(x) \, dt \\ \text{s.t. } \rho(x,0) \equiv \rho_0(x) \;\forall x \in \mathbb{R}^n \\ \rho(x,1) \equiv \rho_1(x) \;\forall x \in \mathbb{R}^n \\ \frac{\partial \rho(x,t)}{\partial t} = -\nabla \cdot (\rho(x,t) v(x,t)) \\ \forall x \in \mathbb{R}^n, t \in (0,1) \end{cases} \tag{16}$$

Here, we assume that we are computing the 2-Wasserstein distance between two distribution functions $\rho_0(x)$ and $\rho_1(x)$. This is often referred to as a *dynamical* model of transport and can be extended to spaces like Riemannian manifolds [**McC01**].

Formulation (16) comes with an intuitive physical interpretation. The time-varying function $\rho(x, t)$ gives the density of a gas as a function of time $t \in [0, 1]$, which starts out in configuration ρ_0 and ends in configuration ρ_1. The constraint $\frac{\partial \rho}{\partial t} = -\nabla \cdot (\rho v)$ is the *continuity equation*, which states that the vector field $v(x, t)$ is the *velocity* of ρ as it moves as a function of time while preserving mass. Over all possible ways to "animate" the motion from ρ_0 to ρ_1, the objective function minimizes $\frac{1}{2}\rho\|v\|_2^2$ (mass times velocity squared): the total kinetic energy!

From a computational perspective, it can be convenient to replace velocity v with momentum $J := \rho \cdot v$ to obtain an equivalent formulation to (16):

$$\mathcal{W}_2^2(\rho_0, \rho_1) = \begin{cases} \min_{J(x,t),\rho(x,t)} \frac{1}{2} \int_0^1 \int_{\mathbb{R}^n} \frac{\|J(x,t)\|_2^2}{\rho(x,t)} \, dA(x) \, dt \\ \text{s.t. } \rho(x,0) \equiv \rho_0(x) \;\forall x \in \mathbb{R}^n \\ \rho(x,1) \equiv \rho_1(x) \;\forall x \in \mathbb{R}^n \\ \frac{\partial \rho(x,t)}{\partial t} = -\nabla \cdot J(x,t) \\ \forall x \in \mathbb{R}^n, t \in (0,1) \end{cases} \tag{17}$$

This formulation is jointly convex in the unknowns (ρ, J).

Dynamical formulations of transport make explicit the phenomenon of *displacement interpolation* [**McC94, McC97**], illustrated in Figure 6. Intuitively, the Wasserstein distance $\mathcal{W}_2$ between two distribution functions ρ_0 and ρ_1 is "explained" by a time-varying sequence of distributions ρ_t interpolating from one to the other. Unlike the trivial interpolation $\rho(t) := (1 - t)\rho_0(x) + t\rho_1(x)$, optimal transport *slides* the distribution across the geometric domain similar to a geodesic shortest path between points on a curved manifold. Indeed, the intuitive connection to differential geometry is more than superficial: [**Ott01**, **Lot08**] show how to interpret (16) as a geodesic in an infinite-dimensional manifold of probability distributions over a fixed domain.

Other p-Wasserstein distances $\mathcal{W}_p$ also admit Eulerian formulations. Most importantly, the 1-Wasserstein distance can be computed as follows:

$$\mathcal{W}_1(\rho_0, \rho_1) = \left\{ \begin{array}{rl} \min_{J(x)} & \int_{\mathbb{R}^n} \|J(x)\|_2 \, dA(x) \\ \text{s.t.} & \nabla \cdot J(x) = \rho_1(x) - \rho_0(x). \end{array} \right. \tag{18}$$

This problem, known as the *Beckmann problem*, has connections to traffic modeling and other tasks in geometry. From a computational perspective, it has the useful property that the vector field $J(x)$ has no time dependence, reducing the number of unknown variables in the optimization problem.

3. Motivating applications

Having developed the basic definition and theoretical properties of the optimal transport problem, we can now divert from theoretical discussion to mention some concrete applications of transport in the computational world. These are just a few, chosen for their diversity (and no doubt biased toward areas adjacent to the author's research); in reality optimal transport is beginning to appear in a huge variety of computational pipelines. Our goal in this section is not to give the details of each problem and its resolution with transport, but just to give a flavor of how optimal transport can be applied as a powerful modeling tool in application-oriented disciplines as well as citations to more detailed treatments of each application.

Operations and logistics. Given its history and even its name, it comes as no surprise that a primary application of optimal transport is in the operations and logistics world, in which engineers are asked to find a minimum-cost routing of packages or materials to customers. The basic theory and algorithms for this case of optimal transport date back to World War II, in which optimal transport of soldiers, weapons, supplies, and the like were by no means theoretical problems.

A particular case of interest in this community is that of transport over a graph $G = (V, E)$. Here, shortest-path distances over the edges of G provide the costs for transport, leading to a problem known to computer scientists as *minimum-cost flow without edge capacities* [**RMO93**]. This linear program is a classic algorithmic problem, with well-known algorithms including cycle canceling [**Kle67**], network simplex [**Orl97**], and the Ford–Fulkerson method [**FJF56**]. A challenge for theoretical computer scientists is to design algorithms achieving lower-bound time complexity for solving this problem; recent progress includes [**She17**], which achieves a near-linear runtime using an approach that almost resembles a numerical algorithm rather than a discrete method.

Histogram-based descriptors. Some of the earliest applications of optimal transport in the computational world come from computer vision [**RTG00**]. Suppose we wish to perform *similarity search* on a database of photographs: Given one photograph, we wish to search the database for other photos that look similar. One reasonable way to do this is to describe each photograph as a histogram—or probability distribution—over the space of colors. Two photographs roughly look similar if they have similar color histograms as measured using optimal transport

FIGURE 7. Level sets of geodesic distance to the front right toe of a 3D camel model approximated using the optimal transport technique [**SRGB14a**].

distances (known in this community as the "Earth Mover's Distance"), giving a simple technique for sorting and searching the dataset.

This basic approach comes up time and time again in the applied world. For images, rather than binning colors into a histogram one could bin the orientations and strengths of the gradients to capture the distribution of edge features [**PW09**]. Recent work has proposed an embedding of the words in an English dictionary into Euclidean space $\mathbb{R}^n$ [**MCCD13**], in which case the words present in a given document become a point cloud or superposition of δ-functions in $\mathbb{R}^n$; application of the Wasserstein ("Word-Movers") distance in this case is an effective technique for document retrieval [**KSKW15**].

Registration. Suppose we wish to use a medical imaging device such as the MRI to track the progress of a neurodegenerative disease. On a regular basis, we might ask the subject to return to the laboratory for a brain scan, each time measuring a signal over the volume of the MRI indicating the presence or absence of brain tissue. These signals can vary drastically from visit to visit, not just due to the progress of the disease but also due to more mundane issues like movement of the patient in the measurement device or nonrigid deformation of the brain itself.

Inspired by issues like those mentioned above, the task of computing a map from one scan to another is known as *registration*, and optimal transport has been proposed time and time again as a tool for this task. The basic idea of these tools is to use the transport map π as a natural way to transfer information from one scan to another [**HZTA04**]. One caveat is worth highlighting: Optimal transport does not penalize splitting mass or making non-elastic deformations in the optimal map, so long as points of mass individually do not move too far. A few recent methods attempt to cope with this final issue, e.g. by combining transport with an elastic deformation method more common in medical imaging [**FCVP17**] or by defining modified versions of optimal transport that are invariant to certain species of deformation [**CG99, Mém11, SPKS16**].

Distance approximation. A predictable property of the p-Wasserstein distance $\mathcal{W}_p$ for distributions over a surface or manifold $\mathcal{M}$ is that the distance between δ-functions centered at two points $x_0, x_1 \in \mathcal{M}$ reproduces the geodesic (shortest-path) distance from x_0 to x_1. While distances in Euclidean space are computable

FIGURE 8. A blue noise pattern generated using [**DGBOD12**] (image courtesy F. de Goes, generated from photograph by F. Durand).

using a closed-form formula, distances along discretized surfaces can be challenging to compute algorithmically, requiring techniques like fast marching [**Set99**], the theoretically-justified but difficult-to-implement MMP algorithm [**MMP87**], or diffusion-based approximation [**CWW13**]. In this regime, fast algorithms for approximating optimal transport distances $\mathcal{W}_p$ restricted to δ-functions actually provide a way to approximate geodesic distances while preserving the triangle inequality [**SRGB14a**]; the level sets of one such approximation are shown in Figure 7.

Blue noise and stippling. Certain laser printers and other devices can only print pages in black-and-white—no gray. The idea of *halftoning* is that gray values between black and white can be approximated in a perceptually reasonable fashion by patterns of black dots of varying radius or location over a white background; the halftoned image can be printed using the black-and-white printer and from a distance appears similar to the original.

A reasonable model for halftoning involves optimal transport. In particular, suppose we think of a grayscale image as a *distribution* of ink on a white page; that is, the image can be understood as a measure $\mu \in \mathrm{Prob}([0,1]^2)$, where $[0,1]^2$ is the unit rectangle representing the image plane. Under the reasonable assumption that ink is conserved, we might attempt to approximate μ with a set of dots of black inks, modeled using δ-functions centered at x_i. The intensity of the dot cannot be modulated (the printer only knows how to print in black-and-white), but the location can be moved, leading to an optimization problem to the effect:

$$\min_{x_1,\ldots,x_n} \mathcal{W}_2^2\left(\mu, \frac{1}{n}\sum_i \delta_{x_i}\right).$$

Here, the variables are the locations of the n dots approximating the image, and the Wasserstein 2-distance is used to measure how well the dots approximate μ. This basic idea is extended in [**DGBOD12**] to a pipeline for computing *blue noise*; an example of their output is shown in Figure 8.

Political redistricting. A few recent attempts to propose political redistricting procedures have incorporated ideas from optimal transport to varying degrees of success. For example, optimal transport might provide one simplistic means of assigning voters to polling centers. The distribution of voters over a map is "transported" to a sparse set of polling places, where distributional constraints reflect the fact that each polling center can only handle so many voters; assigning each voter to his/her closest polling center might cause polling centers in city centers to become overloaded. A few papers have proposed variations on this idea to design compact voting districts e.g. for the US House of Representatives [**SBD07**, **Mil07**, **CAKY17**, **Opt19**]. Many confounding—but incredibly important—factors obscure the application of this simplistic mathematical model in practice, ranging from compliance with civil rights law to the simple decision of a transport cost (e.g. geographical versus road network versus public transportation versus travel time).

Statistical estimation. Parameter estimation is a key task in statistics that involves "explaining" a given dataset using a statistical model. For example, given the set of heights of people in a room $\{h_1, \ldots, h_n\}$, a simple parameter estimation task might be to estimate the mean h_0 and standard deviation σ of a normal (bell curve/Gaussian) distribution $g(h|h_0, \sigma)$ from which the data was likely sampled.

Principal among the techniques for parameter estimation is the *maximum likelihood estimator* (MLE). Continuing in our height data example, assuming the n heights are drawn independently, the joint probability of observing the given set of heights in the room is given by the product

$$P(h_1, \ldots, h_n | h_0, \sigma) = \prod_{i=1}^{n} g(h_i | h_0, \sigma).$$

The MLE of the data is the estimate of (h_0, σ) that maximizes this probability value:

$$(h_0, \sigma)_{\mathrm{MLE}} := \arg\max_{h_0, \sigma} P(h_1, \ldots, h_n | h_0, \sigma).$$

For algebraic reasons it is often easier to maximize the *log likelihood* $\log P(\cdots)$, although this is obviously equivalent to the formulation above.

As an alternative to the MLE, however, the *minimum Kantorovich estimator* (MKE) [**BBR06**] uses machinery from optimal transport. As the name suggests, the MKE estimates the parameters of a distribution by minimizing the transport distance between the parameterized distribution and the empirical distribution from data. For our height problem, the optimization might look like

$$(h_0, \sigma)_{\mathrm{MKE}} := \arg\min_{h_0, \sigma} \mathcal{W}_2^2 \left(\frac{1}{n} \sum_i \delta_{h_i}, g(\cdot | h_0, \sigma) \right)$$

The differences between MLE, MKE, and other alternatives can be subtle from the outside looking in, and the MKE is only recently being studied in applied environments in comparison to more conventional alternatives. Since it takes into account the distance measure of the geometric space on which the samples are defined, the MKE appears to be robust to geometric noise that can confound more traditional alternatives—at the price of increased computational expense. Recent applications have shown value of this estimator for training and inference in machine learning models [**MMC16**, **BJGR17**].

FIGURE 9. Optimal transport is used to design the shape of transparent or reflective material to show a particular caustic pattern (image courtesy of EPFL Computer Graphics and Geometry Laboratory and Rayform SA).

Domain adaptation. Many basic statistical and machine learning algorithms make a false assumption that the "training" and "test" data are distributed equally. As an example where this is not the case, suppose we wish to make an object recognition tool that learns how to label the contents of a photograph. As training data, we use the listings on an e-commerce site like Amazon.com, which contain not only a photograph of a given object but also text describing it. But, while this training data is extremely clean, it is not representative of possible test data, e.g. gathered by a robot navigating a shopping mall: Photographs collected by the latter likely contain clutter, a variety of lighting configurations, and countless other confounding factors. Algorithms designed to compensate for the difference between training and test data are known as *domain adaptation* techniques.

One possibility is to use optimal transport to design a stable domain adaptation tool. The basic idea is to view the training data as a point cloud in some Euclidean space $\mathbb{R}^d$. For instance, perhaps d could be the number of pixels in a photograph; the location of every point in the point cloud determines the contents of the photo, and as additional information each point is labeled with a text name. The test data is also a point cloud in $\mathbb{R}^d$, but thanks to the confounding factors listed above perhaps these two points clouds are not aligned. [**CFTR17**] proposes using optimal transport to align the training data to the test data and to carry the label information along, e.g. attempting to align the space of Amazon.com photos to the space of shopping mall photos. Once the training and test data are aligned, it makes sense to transfer information, classifiers, and the like from one to the other.

Engineering design. Optimal transport has found application in design tools for many engineering tasks, from reflector design [**Oli87, Wan96**] to aerodynamics [**Pla12**]. One intriguing paper uses optimal transport to design transparent objects made of materials like glass, which can focus light into *caustics* via refraction [**STTP14**]. By minimizing the transport distance between the light rays by the glass and a desired black-and-white image, they can "shape" the distribution of light as it comes out of a window. An example caustic design computed using their method is shown in Figure 9.

4. One problem, many discretizations

Computational optimal transport is a relatively new discipline, and techniques for solving the optimal transport problem and in particular computing Wasserstein distances are still a topic of active research. So far, it appears that no "one size fits all" approach has been discovered; rather, different applications and scenarios demand different numerical techniques for optimal transport.

Several desiderata inform the design of an algorithm for optimal transport:

- EFFICIENCY: While L_1 distances and KL divergences are computable using closed-form formulas, optimal transport distance computation requires solving an optimization problem. The cost of solving this problem relative to the cost of direct computation of transport's simpler alternatives is largely the reason why optimal transport has not reached a higher level of popularity in the applied world. But this scenario is changing: New high-speed algorithms for large-scale transport are nearly competitive with more traditional alternatives while bringing to the table the geometric structure unique to transport world.
- STABILITY: A theme in the numerical analysis literature is stability, the resilience of a computation to small changes in the input. Stability of the minimal transport objective value and/or its accompanying transport map can be a challenging topic. Linear program discretizations of continuum optimal transport problems tend to resemble (9) above, a linear program whose optimal solution T *provably* has the sparsity of a permutation matrix; this implies that a small perturbation of v or w may result in a discrete change of T's sparsity structure.
- STRUCTURE PRESERVATION: Transport is well-studied theoretically, and one could reasonably expect that key properties of transport in the infinite-dimensional case are preserved either exactly or approximately when they are computed numerically. For instance, Wasserstein distances enjoy a triangle inequality, and Eulerian formulations of transport have connections to gradient flows and other PDE. Provable guarantees that these structures are preserved in discretizations of transport help assure that nothing critical is lost in the process of approximating transport distances algorithmically.

One reason why there are so many varied algorithms available for numerical OT is that the problem can be written in many different ways (see §2.4). A basic recipe for designing a transport algorithm is to choose any of the equivalent formulations of transport—all of which yield the same optimal value in theory—, discretize any variables that are otherwise infinite-dimensional, and design a bespoke optimization algorithm to solve the resulting problem, which now has a finite number of variables. The flexibility of choosing *which* version of transport to discretize usually is tuned to the geometry of a given application, desired properties of the resulting discretization, and ease of optimizing the discretized problem. The reality of choosing a discretization to facilitate ease of computation reflects a tried-and-true maxim of engineering: "If a problem is difficult to solve, change the problem."

In this section, we roughly outline a few discretizations and accompanying optimization algorithms for numerical OT. Our goal is not to review all well-known techniques for computational transport thoroughly but rather to highlight the breadth of possible approaches and to give a few practical pointers for implementing state-of-the-art transport algorithms at home.

4.1. Regularized transport. We will start by describing *entropically-regularized transport*, a technique that has piqued the interest of the machine learning community after its introduction there in 2013 [**Cut13**]. This technique has an explicit trade-off between accuracy and computational efficiency and has shown particularly strong promise in the regime where a rough estimate of transport is sufficient. This regime aligns well with the demands of "big data" applications, in which individual data points are likely to be noisy—so obtaining an extremely accurate transport value would be overkill computationally.

Regularization is a key technique in optimization and inverse problems in which an objective function is modified to encode additional assumptions and/or to make it easier to minimize. For example, suppose we wish to solve the least-squares problem $\min_x \|Ax - b\|_2^2$ given a matrix $A \in \mathbb{R}^{m\times n}$ and a vector $b \in \mathbb{R}^m$. When A is rank-deficient or if $m < n$, an entire affine space of x's achieve the minimal value. To get around this, we could apply Tikhonov regularization (also known as ridge regression), in which we instead minimize $\|Ax - b\|_2^2 + \alpha\|x\|_2^2$ for some $\alpha > 0$. As $\alpha \to 0$, a solution of the original least-squares problem is recovered, while for any $\alpha > 0$ the regularized problem is guaranteed to have a unique minimizer; as $\alpha \to \infty$, we have $x \to 0$, a predictable but uninteresting value. From a high level, we can think of α as trading off between fidelity to the original problem $Ax \approx b$ and ease of solution: For small $\alpha > 0$ the problem is near-singular but close to the original least-squares formulation, while larger α makes the problem easier to solve.

The variables in the basic formulation of transport are nonnegative probability values, to which it is difficult to apply standard least-squares style Tikhonov regularization; see [**ES18, BSR18**] for discussion of this case. Instead, entropic regularization uses a regularizer from information theory: the entropy of a probability distribution. Suppose a probability measure has distribution function $\rho(x)$. The (differential) entropy of ρ is defined as

$$H[\rho] := -\int \rho(x)\log\rho(x)\,dx. \tag{19}$$

This definition makes two assumptions that are needed to work with entropy, that a probability measure admits a distribution and that it is nonzero everywhere—otherwise $\log\rho(x)$ is undefined. $H[\rho]$ is a concave function of ρ that roughly measures the "fuzziness" of a distribution. Low entropy indicates that a distribution is sharply peaked about a few points, while high entropy indicates that it is more uniformly distributed throughout space.

The basic approach in entropically-regularized transport is to add a small multiple of $-H[\pi]$ to regularize the transport plan π in the OT problem. We will start by discussing the discrete problem (9), which after entropic regularization can be written as follows:

$$\mathrm{OT}_\alpha(v,w;C) := \left\{ \begin{array}{rl} \min_{T\in\mathbb{R}^{k_1\times k_2}} & \sum_{ij} T_{ij}c_{ij} + \alpha\sum_{ij} T_{ij}\log T_{ij} \\ \text{s.t.} & \sum_j T_{ij} = v_i \ \forall i \in \{1,\dots,k_1\} \\ & \sum_i T_{ij} = w_j \ \forall j \in \{1,\dots,k_2\}. \end{array}\right. \tag{20}$$

We are able to drop the $T \geq 0$ constraint because $\log T_{ij}$ in the objective function prevents negative T values.

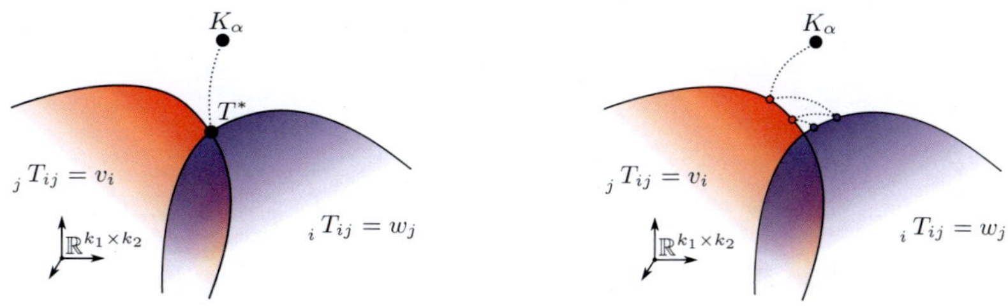

(a) Projection (b) Alternating projection

FIGURE 10. (a) Intuition for the optimization problem (20)as a projection of K_α onto the prescribed row sum and column sum constraints with respect to KL divergence (21). (b) The Sinkhorn algorithm projects back and forth onto one set of constraints and then the other, converging to the transport matrix T^*.

The objective function from (20) can be refactored slightly:

$$
\begin{aligned}
\sum_{ij} T_{ij} c_{ij} + \alpha \sum_{ij} T_{ij} \log T_{ij} &= \alpha \sum_{ij} T_{ij} \left(\frac{c_{ij}}{\alpha} + \log T_{ij} \right) \\
&= \alpha \sum_{ij} T_{ij} \log \frac{T_{ij}}{e^{-c_{ij}/\alpha}} \\
&= \alpha \mathrm{KL}(T | K_\alpha). \qquad (21)
\end{aligned}
$$

Here, we define a *kernel* K_α via $(K_\alpha)_{ij} := e^{-c_{ij}/\alpha}$. KL denotes the Kullback–Leibler divergence [**KL51**], a distance-like (but asymmetric) measure of the similarity between T and K from information theory; the definition of K_α is singular when $\alpha = 0$, indicating that the connection to KL is only possible in the $\alpha > 0$ regime.

(21) gives an intuitive explanation for entropy-regularized transport illustrated in Figure 10(a). The matrix K does not satisfy the constraints of the regularized transport problem (20). Thinking of KL roughly as a distance measure, our job is to find the *closest projection* (with respect to KL) of K onto the set of T's satisfying the constraints $\sum_j T_{ij} = v_i$ and $\sum_i T_{ij} = w_j$. With this picture in mind, Figure 10(b) illustrates the Sinkhorn algorithm for entropy-regularized transport derived below, which alternates between projecting onto one of these sets and then the other.

Continuing in our derivation, we return to (20) to derive first-order optimality conditions. Since (20) is an equality-constrained differentiable minimization problem, it can be solved using a standard multi-variable calculus technique: the method of Lagrange multipliers. There are $k_1 + k_2$ constraints, so we need $k_1 + k_2$ Lagrange multipliers, which—following the derivation of (11)—we store in vectors

$\phi \in \mathbb{R}^{k_1}$ and $\psi \in \mathbb{R}^{k_2}$. The Lagrange multiplier function here is:

$$\begin{aligned}\Lambda(T;\phi,\psi) &:= \sum_{ij} T_{ij}c_{ij} + \alpha\sum_{ij} T_{ij}\log T_{ij}\\ &+\sum_i \phi_i\left(v_i - \sum_j T_{ij}\right) + \sum_j \psi_j\left(w_j - \sum_i T_{ij}\right)\\ &= \langle T, C\rangle + \alpha\langle T, \log T\rangle + \phi^\top(v - T\mathbb{1}) + \psi^\top(w - T^\top\mathbb{1})\end{aligned}$$

Here, $\langle\cdot,\cdot\rangle$ indicates the element-wise inner product of matrices, the log is element-wise, and $\mathbb{1}$ indicates the vector of all ones. Taking the gradient with respect to T gives the following first-order optimality condition:[1]

$$\begin{aligned}&0 = \nabla_T\Lambda = C + \alpha\mathbb{1}\mathbb{1}^\top + \alpha\log T - \phi\mathbb{1}^\top - \mathbb{1}\psi^\top\\ &\implies \log T = \frac{(\phi - \alpha\mathbb{1})\mathbb{1}^\top}{\alpha} + \frac{\mathbb{1}\psi^\top}{\alpha} + \log K_\alpha \text{ where } K_\alpha := \exp[-C/\alpha]\\ &\implies \boxed{T = \operatorname{diag}[p]K_\alpha\operatorname{diag}[q]} \text{ where } p := \exp\left[\frac{\phi - \alpha\mathbb{1}}{\alpha}\right] \text{ and } q := \exp\left[\frac{\psi}{\alpha}\right].\end{aligned}$$

Here, $\operatorname{diag}[v]$ indicates the diagonal matrix whose diagonal is v. The key result is the boxed equation, which gives a formula for the unknown transport matrix T in terms of two unknown vectors p and q derived by changing variables from the Lagrange multipliers ϕ and ψ. There are multiple choices of p and q in terms of ϕ and ψ that all give the same "diagonal rescaling" formula including some that are more symmetric, but this detail is not important.

Next we plug the new relationship $T = \operatorname{diag}[p]K_\alpha\operatorname{diag}[q]$ into the constraints of (20) to find

$$\begin{aligned}p \odot (K_\alpha q) &= v\\ q \odot (K_\alpha^\top p) &= w.\end{aligned} \tag{22}$$

Here, $\odot$ denotes the elementwise (Hadamard) product of two vectors or matrices. These formulas determine the unknown vector p in terms of q and vice versa.

The formulas (22) directly suggest a state-of-the-art technique for entropy-regularized optimal transport, known as the *Sinkhorn (or Sinkhorn–Knopp) algorithm* and dating back to an early technique for matrix rescaling [**SK67**]. This extremely succinct algorithm successively updates estimates of p and q. Iteration k is given by the update formulas ($\oslash$ denotes elementwise division)

$$\begin{aligned}p^{k+1} &\leftarrow v \oslash (K_\alpha q^k)\\ q^{k+1} &\leftarrow w \oslash (K_\alpha^\top p^{k+1}).\end{aligned}$$

It can be implemented in fewer than ten lines of code! The basic approach is to update p in terms of q using the first relationship, then q in terms of p using the second relationship, then p again, and so on. Using essentially the geometric intuition provided in Figure 10(b) for this technique and explored in-depth in [**BCC$^+$15**], one can prove that $\operatorname{diag}[p]K_\alpha\operatorname{diag}[q]$ converges asymptotically to the optimal T at a relatively efficient rate regardless of the initial guess.

[1] Readers uncomfortable with this sort of calculation are strongly encouraged to take a look at the useful "cheat sheet" document [**PP08**].

Several advantages distinguish the Sinkhorn method from its peers. Most critically, beyond its ease of implementation, this algorithm is built from simple linear algebra operations—matrix-vector multiplies and elementwise arithmetic—that parallelize well and can be carried out extremely quickly on modern processing hardware. One modern spin on Sinkhorn shows how to shave off even more calculations while preserving its favorable convergence rate [**AWR17**].

Beyond inspiring a huge body of follow-on work in machine learning and computer vision, the Sinkhorn rescaling algorithm provides a means to adapt optimal transport to discrete domains suggested in [**SDGP$^+$15**]. So far, our description of the Sinkhorn method has been generic to *any* cost matrix C. Adding geometric structure to the problem gives it a strong interpretation using heat flow and suggests a faster way to carry out Sinkhorn iterations on discrete domains.

Suppose that the transport cost C is given by squared pairwise distances along a discretized piece of geometry such as a triangulated surface, denoted Σ; this corresponds to computing a regularized version of the 2-Wasserstein distance (10). The dual variables p and q can be thought of as *functions* over Σ, discretized e.g. using one value per vertex. Then, the kernel K_α has elements

$$(K_\alpha)_{ij} = e^{-d(x_i,x_j)^2/\alpha},$$

where $d(x_i, x_j)$ denotes the shortest-path (geodesic) distance along the domain from vertex i to vertex j.

To start, if our domain is flat, or Euclidean, then $(K_\alpha)_{ij} = e^{-\|x_i-x_j\|_2^2/\alpha}$ for points $\{x_i\}_i \subseteq \mathbb{R}^n$. Considered as a function of the x_i's, we recognize K_α up to scale as a *Gaussian* (or normal distribution, or bell curve) in distance. Multiplication by K_α is then *Gaussian convolution*, an extremely simple operation that can be carried out algorithmically using methods like the Fast Fourier Transform (FFT). In other words, rather than explicitly computing and storing the matrix K_α as an initial step and computing matrix-vector products $K_\alpha p$ and $K_\alpha q$ (note K_α is symmetric in this case) in every iteration of the Sinkhorn algorithm, in this case we can replace these products with convolutions $g_\sigma * p$ and $g_\sigma * q$, where $*$ denotes convolution and g_σ is a Gaussian whose standard deviation is determined by the regularizer α. This is *completely equivalent* to the Sinkhorn method that explicitly computes the matrix-vector product, while eliminating the need to store K_α and improving algorithmic speed thanks to fast Gaussian convolution. Put more simply, in the Euclidean case **multiplication by K_α is more efficient than storing K_α** since we can carry out the former implicitly.

When Σ is curved, we can use a mathematical sleight of hand modifying the entropic regularizer to improve computational properties while maintaining convergence to the true optimal transport value as the regularizer goes to zero. We employ a well-known property of geodesic distances introduced in theory in [**Var67**] and applied to computing distances on discrete domains in [**CWW13**]. This property, known as Varadhan's formula, states that geodesic distance $d(x, y)$ between two points x, y on a manifold can be recovered from heat diffusion over a short time:

$$d(x,y)^2 = \lim_{t\to 0}[-2t \ln \mathcal{H}_t(x,y)].$$

Recall that the heat kernel $\mathcal{H}_t(x, y)$ determines diffusion between $x, y \in \Sigma$ after time t. That is, if f_t satisfies the heat equation $\partial_t f_t = \Delta f_t$, where Δ denotes the

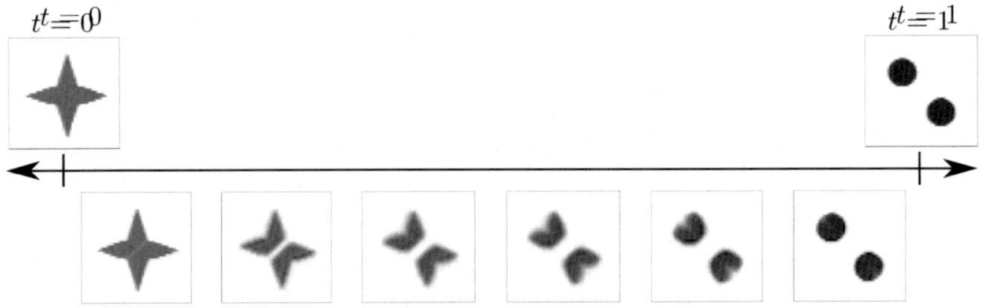

FIGURE 11. Output from an Eulerian algorithm for optimal transport extending [**BB00**] (image courtesy H. Lavenant); interpolation between the two distributions on the top is shown below the timeline. In addition to finding the transport cost, methods in this class also provide a sequence of distributions interpolating between the two inputs.

Laplacian operator, then

$$f_t(x) = \int_\Sigma f_0(y)\mathcal{H}_t(x, y)\, dy.$$

Connecting to the previous paragraph, the heat kernel in Euclidean space is exactly the Gaussian function! Hence, if we replace the kernel K_α with the heat kernel $\mathcal{H}_{\alpha/2}$ in Sinkhorn's method, in the Euclidean case nothing has changed. In the curved case, we get a new approximation of Wasserstein distances introduced as "convolutional Wasserstein distances."

All that remains is to convince ourselves that we can compute matrix-vector products $\mathcal{H}_t \cdot p$ when $\mathcal{H}_t$ is the heat kernel of a discretized domain Σ that is not Euclidean. Thankfully, armed with material from other chapters in this tutorial, this is quite straightforward in the context of discrete differential geometry. In particular, the well-known cotangent approximation of the Laplacian Δ can be combined with standard ordinary differential equation (ODE) solution techniques to carry out heat diffusion in this case using sparse linear algebra. We refer the reader for [**SDGP$^+$15**] for details of one implementation that uses DDG tools extensively.

4.2. Eulerian algorithms. Entropically-regularized transport works with the Kantorovich formulation in (7). This may be one of the earliest and most intuitive definitions of optimal transport, but this in itself is not a strong argument in favor of tackling this formulation numerically. As a point of contrast, we now explore a *completely different* approximation of Wasserstein distances that can be useful in low-dimensional settings, built from the Eulerian (fluid mechanics) formulation of the 2-Wasserstein distance $\mathcal{W}_2^2$ in (16). Historically, this method pre-dates the popularity of entropically-regularized transport and has distinct advanges and disadvantages: It explicitly computes a time-varying displacement interpolation of a density "explaining" the transport (see Figure 11) but in the process must solve a difficult boundary-value PDE problem. Beyond the original paper [**BB00**], we recommend the excellent tutorial [**Pey10**] that steps through an implementation of this technique in practice.

We make a few more simplifications to the continuum formulation before discretizing it. We start by making a quick observation: for any vector $J \in \mathbb{R}^n$ and $\rho > 0$ we have

$$\frac{\|J\|_2^2}{2\rho} = \left\{ \begin{array}{rl} \sup_{a\in\mathbb{R},b\in\mathbb{R}^n} & a\rho + b^\top J \\ \text{s.t.} & a + \frac{\|b\|_2^2}{2} \leq 0. \end{array} \right.$$

This convex program not only justifies that the quotient $\|J\|_2^2/2\rho$ is convex jointly in J and ρ, but also it shows we can write the optimization problem (17) with additional variables as

$$\begin{array}{rl} \inf_{J,\rho}\sup_{a,b} & \int_0^1 \int_{\mathbb{R}^n} [a(x,t)\rho(x,t) + b(x,t)^\top J(x,t)]\, dA(x)\, dt \\ \text{s.t.} & \rho(x,0) \equiv \rho_0(x)\ \forall x \in \mathbb{R}^n \\ & \rho(x,1) \equiv \rho_1(x)\ \forall x \in \mathbb{R}^n \\ & \frac{\partial \rho(x,t)}{\partial t} = -\nabla \cdot J(x,t)\ \forall x \in \mathbb{R}^n, t\in(0,1) \\ & a(x,t) + \frac{\|b(x,t)\|_2^2}{2} \leq 0\ \forall x \in \mathbb{R}^n, t \in (0,1). \end{array}$$

Next, we introduce a dual potential function $\phi(x,t)$ similarly to the derivation of (12) to take care of all but the last constraint:

$$\begin{array}{rl} \inf_{J,\rho}\sup_{a,b,\phi} & \int_0^1 \int_{\mathbb{R}^n} \Big[a(x,t)\rho(x,t) + b(x,t)^\top J(x,t) \\ & +\phi(x,t)\left(\frac{\partial\rho(x,t)}{\partial t} + \nabla\cdot J(x,t)\right)\Big]\, dA(x)\, dt \\ & + \int_{\mathbb{R}^n} [\phi(x,1)(\rho_1(x) - \rho(x,1)) - \phi(x,0)(\rho_0(x) - \rho(x,0))]\, dA(x) \\ \text{s.t.} & a(x,t) + \frac{\|b(x,t)\|_2^2}{2} \leq 0\ \forall x\in\mathbb{R}^n, t\in(0,1). \end{array} \tag{23}$$

We can simplify some terms in this expression. First, using integration by parts we have

$$\int_0^1 \phi(x,t)\frac{\partial \rho(x,t)}{\partial t}\, dt = [\rho(x,1)\phi(x,1) - \rho(x,0)\phi(x,0)] - \int_0^1 \rho(x,t)\frac{\partial\phi(x,t)}{\partial t}\, dt$$

We also can integrate by parts in x to show

$$\int_{\mathbb{R}^n} \phi(x,t)\nabla\cdot J(x,t)\, dA(x) = -\int_{\mathbb{R}^n} J(x,t)^\top \nabla\phi(x,t)\, dA(x).$$

This simplification works equally well if we replace $\mathbb{R}^n$ with the box $[0,1]^n$ with periodic boundary conditions. Incorporating these two integration by parts formulae into our objective function yields a new one:

$$\int_{\mathbb{R}^n} \left\{ \int_0^1 \left(\rho(x,t)\left[a(x,t) - \frac{\partial\phi(x,t)}{\partial t}\right] + J(x,t)^\top[b(x,t) - \nabla\phi(x,t)]\right) dt \right.$$
$$\left. - \phi(x,0)\rho_0(x) + \phi(x,1)\rho_1(x)) \right\} dA(x)$$

We now make some notational simplifications. Define $z := \{\rho, J\}$ and $q := \{a, b\}$ with inner product

$$\langle z, q\rangle := \int_{\mathbb{R}^n}\int_0^1 (a(x,t)\rho(x,t) + b(x,t)^\top J(x,t))\, dt\, dA(x).$$

Furthermore, define

$$F(q) := \begin{cases} 0 & \text{if } a(x,t) + \frac{\|b(x,t)\|_2^2}{2} \leq 0 \ \forall x \in \mathbb{R}^n, t \in (0,1) \\ \infty & \text{otherwise.} \end{cases}$$

$$G(\phi) := \int_{\mathbb{R}^n} (\phi(x,0)\rho_0(x) - \phi(x,1)\rho_1(x))\, dA(x)$$

These functions are both convex. These functions, plus our simplifications and a sign change, allow us to write (23) in a compact fashion as:

$$-\sup_z \inf_{q,\phi} \left[F(q) + G(\phi) + \langle z, \nabla_{x,t}\phi - q\rangle\right], \tag{24}$$

where $\nabla_{x,t}\phi := \{\partial\phi/\partial t, \nabla_x \phi\}$.

Blithely assuming strong duality, namely that we can swap the supremum and the infimum, we arrive at an alternative interpretation of (24). In particular, we can view z as a Lagrange multiplier corresponding to a constraint $q = \nabla_{x,t}\phi$. From this perspective, we actually can find a saddle point (max in z, minimum in (q, ϕ)) of the *augmented Lagrangian* L_r for any $r \geq 0$:

$$L_r(\phi, q, z) := F(q) + G(\phi) + \langle z, \nabla_{x,t}\phi - q\rangle + \frac{r}{2}\langle \nabla_{x,t}\phi - q, \nabla_{x,t}\phi - q\rangle.$$

The extra term here effectively adds zero to the objective function, assuming the constraint is satisfied.

The algorithm proposed in [**BB00**] iteratively updates estimates $(\phi^\ell, q^\ell, z^\ell)$ by cycling through the following three steps:

$$\begin{aligned} \phi^{\ell+1} &\leftarrow \arg\min_\phi L_r(\phi, q^\ell, z^\ell) \\ q^{\ell+1} &\leftarrow \arg\min_q L_r(\phi^{\ell+1}, q, z^\ell) \\ z^{\ell+1} &\leftarrow z^\ell + r(q^{\ell+1} - \nabla_{x,t}\phi^{\ell+1}). \end{aligned}$$

The first two steps update some variables while holding the rest fixed to the best possible value. The third step is gradient step for z. This cycling algorithm and equivalent formulations has many names in the literature—including ADMM [**BPC**$^+$**11**], the Douglas–Rachford algorithm [**DR56, LM79**], and the Uzawa algorithm [**Uza68**]—and is known to converge under weak assumptions.

The advantage of this algorithm is that the individual update formulae are straightforward. In particular, the ϕ update is equivalent to solving a Laplace equation

$$\Delta_{x,t}\phi^{\ell+1} = \nabla_{x,t} \cdot (z^\ell - rq^\ell),$$

where $\Delta_{x,t}$ is the Laplacian operator in time and space. The q update decouples over x and t, amounting to projecting $\nabla_{x,t}\phi^{\ell+1} + z^\ell/r$ onto the constraints in the definition of $F(q)$ with respect to L_2, a one-dimensional problem solvable analytically. And, the z update is already in closed-form.

So far, we have described the Benamou–Brenier algorithm using continuum variables, but of course at the end of the day we must discretize the problem for computational purposes. The most straightforward discretization assumes ρ_0 and ρ_1 are supported in the unit square $[0,1]^n$, which is broken up into a $m \times m \times \cdots \times m$ grid, and further discretizes the time variable $t \in [0,1]$ into p steps. Then, all degrees of freedom (ϕ, q, z) can be put on the grid vertices and interpolated in between using multilinear basis functions; this leads to a finite element (FEM) discretization of the problem that can be approached using techniques discussed in earlier chapters. An alternative grid-based discretization and accompanying optimization algorithm is also given in [**PPO14**].

The use of PDE language makes this dynamical formulation of transport potentially compatible with machinery like discrete exterior calculus (DEC) [**Hir03**], which could be used to define a discrete notion of transport on simplicial complexes like triangle meshes. [**LCCS18**] provides one such notion of "discrete dynamical transport," discretizing the spatial variables onto a mesh but leaving the continuum interval of time t alone. In their language, distributions on a mesh are interpreted as a finite-dimensional manifold parameterized by one density value per vertex and equipped with a careful choice of Riemannian metric; transport distances become geodesic distances along this manifold. Generally speaking, discretizing the objective function $\|J\|_2^2/\rho$ on a triangle mesh is challenging because scalar quantities like ρ typically are discretized on vertices or faces while vectorial quantities like J are better suited for edges. Evaluating $\|J\|^2/2\rho$ then requires averaging J or ρ so that the two end up on the same simplices. Some recent papers with analogous constructions on graphs [**Maa11, SRGB16, ERSS17**] consider related challenges on one-dimensional structures.

While the Benamou–Brenier dynamical formulation of transport is the best known, it is worth noting that the Beckmann problem (18) for the 1-Wasserstein distance $\mathcal{W}_1$ more readily admits discretization using the finite element method (FEM) while preserving a triangle inequality. Details of such a formulation as well as an efficient optimization algorithm are provided in [**SRGB14a**]. The reason (18) is easier to discretize is that the time-varying aspect of transport is lost in this formulation: All that is needed is a single vector $J(x)$ per point x. What makes this problem easy to discretize and optimize is its downfall application-wise: Interpolation with respect to $\mathcal{W}_1$ between two densities μ_0 and μ_1 is given by the uninteresting solution $\mu_t = (1-t)\mu_0 + t\mu_1$, which does not displace mass but rather "teleports" it from the source to the target.

Another PDE-based approach to optimal transport is worth noting and has strong connections to the theory of transport without connecting to fluid flow. Recall the Monge formulation of optimal transport on $\mathbb{R}^n$ in (3), which seeks a map $\phi(x)$ that pushes forward one distribution function $\rho_0(x)$ onto another $\rho_1(x)$. A famous result by Brenier [**Bre91**] shows that ϕ can be written as the gradient of a convex potential $\Psi(x)$: $\phi(x) = \nabla\Psi(x)$. Using H to denote the Hessian operator, this potential satisfies the Monge-Ampère PDE

$$\det(H\Psi(x))\rho_1(\nabla\Psi(x)) = \rho_0(x), \tag{25}$$

a second-order nonlinear elliptic equation that is extremely challenging to solve in practice. A few algorithms, e.g. [**OP89, LR05, BFO10, FO11, BFO14**], tackle this nonlinear system head-on, discretizing the variables involved and solving for Ψ.

4.3. Semidiscrete transport. Our final example from the computational transport world uses yet another formulation of the transport problem. This time, our inspiration is the one-dimensional semidiscrete problem, whose solution is motivated from the formulation in (6). Our exposition of this problem closely follows the excellent tutorial [**LS18**].

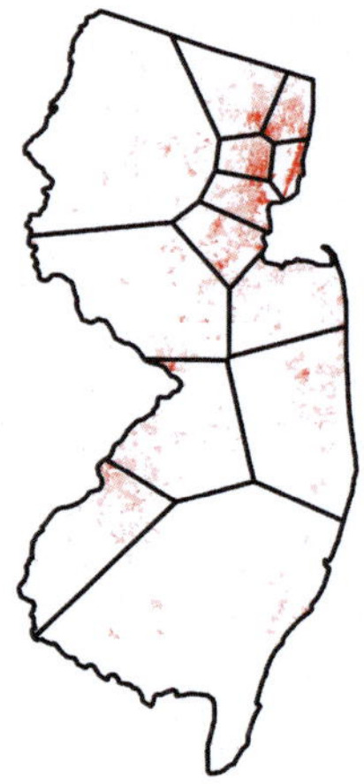

FIGURE 12. Power diagram from a semidiscrete transport problem (image courtesy R. Barnes). Here, semidiscrete transport is used to partition the state of New Jersey into cells with equal population; population density is shaded in red.

In this setting, optimal transport is computed from a distribution whose mass is concentrated at a finite set of isolated points to a distribution with a known but potentially smooth density function. Recall that in the one-dimensional case, we learned that each point of mass in the source is mapped to an *interval* in the target. That is, the domain of the target density is partitioned into contiguous cells whose mass is assigned to a single source point. We will find that the higher-dimensional analog is spiritually identical: Each point of mass in the source density is assigned to a convex region of space in the target. This observation will suggest algorithms constructed from ideas in discrete geometry extending Voronoi diagrams and similar constructions.

As in (6), suppose we are computing the 2-Wasserstein distance from a discrete measure $\mu := \sum_{i=1}^k a_i \delta_{x_i}$, whose mass is concentrated at points $x_i \in \mathbb{R}^n$ with weights $a_i > 0$, to an absolutely continuous measure ν with distribution function $\rho(x)$. The dual formulation of transport (12) in this case can be written

$$\begin{array}{rl} \sup_{\phi,\psi} & \sum_{i=1}^k a_i \phi(x_i) + \int_{\mathbb{R}^n} \psi(y)\rho(y)\, dA(y) \\ \text{s.t.} & \phi(x) + \psi(y) \leq c(x,y)\ \forall x, y \in \mathbb{R}^n. \end{array}$$

The objective in this case "does not care" about values of $\phi(x)$ for $x \notin \{x_i\}_{i=1}^k$. Define $\phi_i := \phi(x_i)$. By this observation, we can write a problem with only one continuum variable:

$$\begin{array}{rl} \sup_{\phi,\psi} & \sum_i a_i \phi_i + \int_{\mathbb{R}^n} \psi(y)\rho(y)\, dA(y) \\ \text{s.t.} & \phi_i + \psi(y) \leq c(x_i,y)\ \forall y \in \mathbb{R}^n, i \in \{1,\ldots,k\}. \end{array}$$

In a slight abuse of notation, for the rest of this section we will think of ϕ as a vector $\phi \in \mathbb{R}^k$ rather than a function $\phi(x)$. Given the supremum, we might as well choose the largest ψ possible that satisfies the constraints. Hence,

$$\psi(y) = \inf_{i \in \{1,\ldots,k\}} [c(x_i, y) - \phi_i].$$

This leads to a final optimization problem in a *finite* set of variables $\phi_1, \ldots, \phi_k$:

$$\begin{aligned}\mathcal{W}_2^2(\mu,\nu) &= \sup_{\phi\in\mathbb{R}^k} \sum_i a_i\phi_i + \int_{\mathbb{R}^n} \rho(y)\left(\inf_{i\in\{1,\ldots,k\}} [c(x_i,y)-\phi_i]\right) dA(y) \\ &= \sup_{\phi\in\mathbb{R}^k} \sum_i \left[a_i\phi_i + \int_{\mathrm{Lag}_\phi^c(x_i)} \rho(y)[c(x_i,y)-\phi_i]\, dA(y)\right] \qquad (26)\end{aligned}$$

Here, $\mathrm{Lag}_\phi^c(x_i)$ indicates the *Laguerre cell* corresponding to x_i:

$$\mathrm{Lag}_\phi^c(x_i) := \{y \in \mathbb{R}^n : c(x_i,y) - \phi_i \leq c(x_j,y) - \phi_j \ \forall j \neq i\}. \qquad (27)$$

The set of Laguerre cells yields the *Laguerre diagram*, a partition of $\mathbb{R}^n$ determined by the cost function c and the vector ϕ; the ϕ_i's effectively control the sizes of the Laguerre cells in the diagram. When $c(x,y) = \|x-y\|_2$ is a distance function and $\phi = 0$, the Laguerre diagram equals the well-known Voronoi diagram of the x_i's that partitions $\mathbb{R}^n$ into loci of points S_i corresponding to those closer to x_i than to the other x_j's [**Aur91**]. More importantly for the 2-Wasserstein distance, when $c(x,y) = 1/2\|x-y\|_2^2$, the Laguerre diagram is known as the *power diagram*, an object studied since the early days of computational geometry [**Aur87**]; an example is shown in Figure 12.

Since (26) comes from a simplification of the dual of the transport problem, it is concave in ϕ; a direct proof can be found in [**AHA92**]. This implies that a simple gradient ascent procedure starting from any initial estimate of ϕ will reach a global optimum. Define the objective function

$$F(\phi) := \sum_i \left[a_i\phi_i + \int_{\mathrm{Lag}_\phi^c(x_i)} \rho(y)[c(x_i,y)-\phi_i]\, dA(y)\right].$$

The gradient can be computed using the partial derivative expression

$$\frac{\partial F}{\partial \phi_i} = a_i - \int_{\mathrm{Lag}_\phi^c(x_i)} \rho(y)\, dA(y). \qquad (28)$$

This expression is predictable from the definition of $F(\phi)$ but must be checked using Reynolds' transport theorem or an analogous result; a similar formula exists for the second derivatives of F. Setting the gradient (28) to zero formalizes an intuition for the optimization problem (26), that it resizes the Laguerre cells by modifying the ϕ_i's until the cell corresponding to each x_i contains mass a_i:

$$a_i = \int_{\mathrm{Lag}_\phi^c(x_i)} \rho(y)\, dA(y).$$

The main ingredient needed to compute the derivatives of F is an algorithm for integrating ρ over Laguerre cells. Hence, gradient ascent and Newton's method applied to optimizing for ϕ cycle between updating the Laguerre diagram for the current ϕ estimate, recomputing the gradient and/or Hessian, assembling these into a search direction, and updating the current estimate of ϕ. For squared Euclidean costs, these algorithms are facilitated by fast algorithms for computing power diagrams, e.g. [**Bow81, Wat81**]. While convergence of gradient descent with line search follows directly from concavity, [**KMT16**] proves that under certain assumptions a damped version of Newton's algorithm—which employs the Hessian in addition to the gradient to accelerate convergence—exhibits global convergence.

Example techniques following this template include [**CGS10**], which proposed an early technique for 2D problems; [**Mér11**], for semi-discrete transport to piece-wise-linear distribution functions in 2D supported on triangle meshes improved using a multiscale approximation; and [**Lév15**], which proposes semi-discrete transport to distributions in 3D that are piecewise-linear on tetrahedral meshes. [**DGBOD12**] provides an early example of a Newton solver for 2D semidiscrete transport using power diagrams and additionally uses derivatives of transport in the support points x_i and weights a_i for assorted approximation problems.

Beyond providing fast algorithms for transport in the semidiscrete case, this formulation is valuable for applications incorporating transport terms. [**DGCSAD11**] employs semidiscrete transport to a collection of distributions concentrated on line segments to reconstruct line drawings from point samples; [**DCSA$^+$14**] proposes a similar technique for reconstructing triangulated surfaces from point clouds in $\mathbb{R}^3$. [**GMMD14**] defines a version of semi-discrete transport intrinsic to a triangulated surface, which can be used for tasks like parameterizing the set of per-vertex area weights in terms of the values ϕ_i.

5. Beyond transport

Beyond improving tools for solving the basic optimal transport problem, some of the most exciting recent work in computational transport involves using transport as a single term in a larger model. In a recent tutorial for the machine learning community, we termed this new trend *"Wassersteinization"* [**CS17**]: using Wasserstein distances to improve geometric properties of variational models in statistics, learning, applied geometry, and other disciplines. Further extending the scope of applied transport, variations of the basic problem have been proposed to apply OT to objects other than probability distributions.

While a complete survey of these creative new applications and extensions is far beyond the scope of this tutorial, we highlight a few interesting pointers into the literature:

- UNBALANCED TRANSPORT: One limitation of the basic model for optimal transport is that it is a distance between histograms or probability distributions, rather than a distance between functions or vectors in $\mathbb{R}^n$—which may not integrate to 1 or may contain negative values. This leads to the problem of *unbalanced transport*, in which mass conservation and/or positivity must be relaxed. Models for this problem range from augmenting the transport problem with a "trash can" that can add or remove mass from distributions [**PW09**] to extensions of dynamical transport to this case [**CPSV16**]. Making transport work for functions rather than distributions while preserving the triangle inequality and other basic properties is challenging both theoretically and from a numerical perspective.
- BARYCENTERS: The idea of displacement interpolation we motivated using (17) suggests a generalization to more than two distributions, known as the *Wasserstein barycenter* problem [**AC11**]. Given k distributions $\mu_1, \ldots, \mu_k$, the Wasserstein barycenter $\mu_{\text{barycenter}}$ is defined as the minimizer of the following optimization problem

$$\mu_{\text{barycenter}} := \arg\min_{\mu} \sum_{i=1}^{k} \mathcal{W}_2^2(\mu, \mu_i). \tag{29}$$

The Wasserstein barycenter gives some notion of *averaging* a set of probability distributions, motivated by observing that the average $\frac{1}{k}\sum_{i=1}^{k} x_i$ of a set of vectors $x_i \in \mathbb{R}^n$ is the minimizer $\arg\min_x \sum_i \|x - x_i\|_2^2$. Barycenter algorithms range from extensions of the Sinkhorn algorithm [**BCC$^+$15, SDGP$^+$15**] to methods that perform gradient descent on μ by differentiating the distance $\mathcal{W}_2$ in its argument [**CD14**] and stochastic techniques requiring only samples from the distributions μ_i [**SCSJ17, CCS18**]. Other algorithms are inspired by a connection to multi-marginal transport [**Pas15**], a generalization of optimal transport involving a distribution over the product of more than two measures. The optimization problem (29) is also one of the earliest examples of "Wassersteinization," in the sense that it is an optimization problem for an unknown distribution μ including Wasserstein distance terms, contrasting somewhat from the optimization problems we considered in §4 in which the unknown is the transport distance itself.

Further generalizing the barycenter problem leads to a notion of the Dirichlet energy of a map from points in one space to distributions over another [**Bre03, Lav17**], with applications in machine learning [**SRGB14b**] and shape matching [**SGB13, MCSK$^+$17**]. An intriguing recent paper also proposes an inverse problem for barycentric coordinates seeking weights for (29) that "explain" an input distribution as a transport barycenter of others [**BPC16**].

- QUADRATIC ASSIGNMENT: The basic optimization problem for transport has an objective function that is *linear* in the unknown transport matrix, expressing a preference for transport maps that do not move any single particle of probabilistic mass very far. This model, however, does not necessarily extract *smooth* maps, wherein distance traveled by any single particle is less important than making sure that nearby particles in the source are mapped to nearby locations in the target. Such a smoothness term leads to a *quadratic* term in the transport problem and allows it to be extended to a distance between metric-measure spaces known as the Gromov–Wasserstein distance [**Mém11, Mém14**], inspired by the better-known but more rigid Gromov–Hausdorff distance. From an optimization perspective, Gromov–Wasserstein computation leads to a "quadratic assignment" problem, known in the most general case to be NP-hard [**SG76**]; practical instances of the problem, however, can be tackled using spectral [**Mém09**] or entropy-based [**SPKS16**] approximations and have shown promise for applications in shape matching [**SPKS16**] and word translation [**AMJ18**]. [**PCS16**] proposes a method for averaging metric spaces using a barycenter formulation similar to (29).
- CAPACITY-CONSTRAINED TRANSPORT: Another extension of the transport problem comes from introducing *capacity constraints* limiting the amount of mass that can travel between assorted pairs of source and target points; in the measure-theoretic formulation, this amounts to constraining transport plan to be dominated by another input plan [**KM15**]. This constraint makes sense in many operations-type applications and has intriguing theoretical properties, but design of algorithms and discretizations

for capacity-constrained transport remains largely open although [**BB00**] provides one approach again extending Sinkhorn's algorithm.

- GRADIENT FLOWS AND PDE: Given a function $f : M \to \mathbb{R}$ defined over a geometric space M like a manifold, a *gradient flow* of f starting at some $x_0 \in M$ attempts to minimize f via "gradient descent" from $x(0) := x_0$ expressed as an ordinary differential equation (ODE) $x'(t) = -\nabla f(x(t))$. Since OT puts a geometry on the space of distributions $\text{Prob}(\mathbb{R}^n)$ over $\mathbb{R}^n$, we can define an analogous procedure that flows probability distributions to reduce certain functionals [**JKO98, San17**]. For instance, gradient flow on the entropy functional (19) in the Wasserstein metric leads to the heat diffusion equation $\partial\rho/\partial t = -\Delta\rho$, where Δ is the Laplacian operator; that is, performing gradient descent on entropy in the Wasserstein metric is exactly the same as diffusing the initial probability distribution like an unevenly-heated metal plate. Beyond giving a variational motivation for certain PDE, this mathematical idea inspired numerical methods for solving PDE that can be written as gradient flows [**Pey15, BCL16**]. Recent work has even incorporated transport into numerical methods for PDE that cannot easily be written as gradient flows in Wasserstein space, such as those governing incompressible fluid flow [**LHO10, Mir15, dGWH$^+$15, MM16**]. Gradient flow properties can also be leveraged as structure to be preserved in discrete models of transport; for instance, [**Maa11**] proposes a model for dynamical optimal transport on a graph and checks that the gradient flow of entropy—now an ODE rather than a PDE—agrees with a discrete heat equation.
- MATRIX FIELDS AND VECTOR MEASURES: *Vector measures* generalize probability measures by replacing scalar-valued probability values $\mu(S) \in [0, 1]$ with values in other cones $\mathcal{C}$. For instance, a tensor-valued measure μ assigns measurable sets S to $d \times d$ postive semidefinite matrices $\mu(S) \in \mathcal{S}_+^d$ while satisfying analogous axioms to those laid out for probability measures in §2.1. These tensor fields find application in diffusion tensor imaging (DTI), which measures diffusivity of molecules like water in the interior of the human brain as a proxy for directionality of white matter fibers; OT extended to this setting can be used to align multiple such images. A few recent models extend OT to this case and propose related numerical methods [**NGT15, CGT17, PCVS17**].

6. Conclusion

The techniques covered in this tutorial are just a few of many ways to approach discrete optimal transport. New algorithms are proposed every month, and there is considerable room for mathematical, algorithmic, and application-oriented researchers to improve existing methods or make their own for different types of data or geometry. Furthermore, mathematical properties such as convergence and approximation quality are still being established for new techniques. Many questions also remain in linking to other branches of discrete differential geometry, e.g. at the most fundamental level defining a *purely discrete* notion of optimal transport compatible with polyhedral meshes or simplicial complexes without requiring regularization and while preserving as much structure as possible from the smooth case.

These challenges aside, discrete optimal transport is demonstrating that OT holds interest far beyond mathematical analysis. New discretizations and algorithms bring down OT's complexity to the point where it can be incorporated into practical engineering pipelines and into larger models without incurring a huge computational expense. Further research into this new discipline holds unique potential to improve both theory and practice and eventually to bring insight into other branches of discrete and smooth geometry.

Acknowledgments

Many thanks to MIT Geometric Data Processing Group members Mikhail Bessmeltsev, Edward Chien, Sebastian Claici, David Palmer, and Dima Smirnov for proofreading this chapter.

References

[AC11] Martial Agueh and Guillaume Carlier, *Barycenters in the Wasserstein space*, SIAM J. Math. Anal. **43** (2011), no. 2, 904–924, DOI 10.1137/100805741. MR2801182

[ACB17] Martin Arjovsky, Soumith Chintala, and Léon Bottou, *Wasserstein generative adversarial networks*, International Conference on Machine Learning, 2017, pp. 214–223.

[AHA92] Franz Aurenhammer, Friedrich Hoffmann, and Boris Aronov, *Minkowski-type theorems and least-squares partitioning*, Proceedings of the Eighth Annual Symposium on Computational Geometry, ACM, 1992, pp. 350–357.

[AMJ18] David Alvarez-Melis and Tommi Jaakkola, *Gromov–Wasserstein alignment of word embedding spaces*, Conference on Empirical Methods in Natural Language Processing, 2018, pp. 1881–1890.

[Aur87] F. Aurenhammer, *Power diagrams: properties, algorithms and applications*, SIAM J. Comput. **16** (1987), no. 1, 78–96, DOI 10.1137/0216006. MR873251

[Aur91] ———, *Voronoi diagrams—a survey of a fundamental geometric data structure*, ACM Computing Surveys (CSUR) **23** (1991), no. 3, 345–405.

[AWR17] Jason Altschuler, Jonathan Weed, and Philippe Rigollet, *Near-linear time approximation algorithms for optimal transport via Sinkhorn iteration*, Proc. NIPS, 2017, pp. 1961–1971.

[BB00] Jean-David Benamou and Yann Brenier, *A computational fluid mechanics solution to the Monge-Kantorovich mass transfer problem*, Numer. Math. **84** (2000), no. 3, 375–393, DOI 10.1007/s002110050002. MR1738163

[BBR06] Federico Bassetti, Antonella Bodini, and Eugenio Regazzini, *On minimum Kantorovich distance estimators*, Statist. Probab. Lett. **76** (2006), no. 12, 1298–1302, DOI 10.1016/j.spl.2006.02.001. MR2269358

[BCC+15] Jean-David Benamou, Guillaume Carlier, Marco Cuturi, Luca Nenna, and Gabriel Peyré, *Iterative Bregman projections for regularized transportation problems*, SIAM J. Sci. Comput. **37** (2015), no. 2, A1111–A1138, DOI 10.1137/141000439. MR3340204

[BCL16] Jean-David Benamou, Guillaume Carlier, and Maxime Laborde, *An augmented Lagrangian approach to Wasserstein gradient flows and applications* (English, with English and French summaries), Gradient flows: from theory to application, ESAIM Proc. Surveys, vol. 54, EDP Sci., Les Ulis, 2016, pp. 1–17, DOI 10.1051/proc/201654001. MR3565819

[BFO10] Jean-David Benamou, Brittany D. Froese, and Adam M. Oberman, *Two numerical methods for the elliptic Monge-Ampère equation*, M2AN Math. Model. Numer. Anal. **44** (2010), no. 4, 737–758, DOI 10.1051/m2an/2010017. MR2683581

[BFO14] Jean-David Benamou, Brittany D. Froese, and Adam M. Oberman, *Numerical solution of the optimal transportation problem using the Monge-Ampère equation*, J. Comput. Phys. **260** (2014), 107–126, DOI 10.1016/j.jcp.2013.12.015. MR3151832

[BJGR17] Espen Bernton, Pierre E Jacob, Mathieu Gerber, and Christian P Robert, *Inference in generative models using the Wasserstein distance*, arXiv:1701.05146 (2017).

[Bow81] A. Bowyer, *Computing Dirichlet tessellations*, Comput. J. **24** (1981), no. 2, 162–166, DOI 10.1093/comjnl/24.2.162. MR619576

[BPC+11] Stephen Boyd, Neal Parikh, Eric Chu, Borja Peleato, and Jonathan Eckstein, *Distributed optimization and statistical learning via the alternating direction method of multipliers*, Foundations and Trends in Machine Learning **3** (2011), no. 1, 1–122.

[BPC16] Nicolas Bonneel, Gabriel Peyré, and Marco Cuturi, *Wasserstein barycentric coordinates: histogram regression using optimal transport*, ACM Transactions on Graphics **35** (2016), no. 4, 71–1.

[Bre91] Yann Brenier, *Polar factorization and monotone rearrangement of vector-valued functions*, Comm. Pure Appl. Math. **44** (1991), no. 4, 375–417, DOI 10.1002/cpa.3160440402. MR1100809

[Bre03] ———, *Extended Monge–Kantorovich theory*, Lecture Notes in Mathematics (2003), 91–122.

[BSR18] Mathieu Blondel, Vivien Seguy, and Antoine Rolet, *Smooth and sparse optimal transport*, International Conference on Artificial Intelligence and Statistics, 2018, pp. 880–889.

[BV04] Stephen Boyd and Lieven Vandenberghe, *Convex optimization*, Cambridge University Press, Cambridge, 2004. MR2061575

[CA14] Marco Cuturi and David Avis, *Ground metric learning*, J. Mach. Learn. Res. **15** (2014), 533–564. MR3190849

[CAKY17] Vincent Cohen-Addad, Philip N Klein, and Neal E Young, *Balanced power diagrams for redistricting*, arXiv:1710.03358 (2017).

[CCS18] Sebastian Claici, Edward Chien, and Justin Solomon, *Stochastic Wasserstein barycenters*, International Conference on Machine Learning, 2018, pp. 998–1007.

[CD14] Marco Cuturi and Arnaud Doucet, *Fast computation of Wasserstein barycenters*, International Conference on Machine Learning, 2014, pp. 685–693.

[CFTR17] Nicolas Courty, Rémi Flamary, Devis Tuia, and Alain Rakotomamonjy, *Optimal transport for domain adaptation*, PAMI **39** (2017), no. 9, 1853–1865.

[CG99] Scott Cohen and Leonidas Guibas, *The earth mover's distance under transformation sets*, Proc. ICCV, vol. 2, IEEE, 1999, pp. 1076–1083.

[CGS10] G. Carlier, A. Galichon, and F. Santambrogio, *From Knothe's transport to Brenier's map and a continuation method for optimal transport*, SIAM J. Math. Anal. **41** (2009/10), no. 6, 2554–2576, DOI 10.1137/080740647. MR2607321

[CGT17] Yongxin Chen, Tryphon T. Georgiou, and Allen Tannenbaum, *Matrix optimal mass transport: a quantum mechanical approach*, IEEE Trans. Automat. Control **63** (2018), no. 8, 2612–2619, DOI 10.1109/tac.2017.2767707. MR3845989

[Coh80] Donald L. Cohn, *Measure theory*, Birkhäuser, Boston, Mass., 1980. MR578344

[CPSV16] Lénaïc Chizat, Gabriel Peyré, Bernhard Schmitzer, and François-Xavier Vialard, *An interpolating distance between optimal transport and Fisher-Rao metrics*, Found. Comput. Math. **18** (2018), no. 1, 1–44, DOI 10.1007/s10208-016-9331-y. MR3749413

[CS17] Marco Cuturi and Justin Solomon, *A primer on optimal transport*, NIPS Tutorial, 2017.

[Cut13] Marco Cuturi, *Sinkhorn distances: Lightspeed computation of optimal transport*, Advances in Neural Information Processing Systems, 2013, pp. 2292–2300.

[CWW13] Keenan Crane, Clarisse Weischedel, and Max Wardetzky, *Geodesics in heat: A new approach to computing distance based on heat flow*, ACM Transactions on Graphics (TOG) **32** (2013), no. 5, 152.

[DCSA+14] Julie Digne, David Cohen-Steiner, Pierre Alliez, Fernando de Goes, and Mathieu Desbrun, *Feature-preserving surface reconstruction and simplification from defect-laden point sets*, J. Math. Imaging Vision **48** (2014), no. 2, 369–382, DOI 10.1007/s10851-013-0414-y. MR3152110

[DGBOD12] Fernando De Goes, Katherine Breeden, Victor Ostromoukhov, and Mathieu Desbrun, *Blue noise through optimal transport*, ACM Transactions on Graphics (TOG) **31** (2012), no. 6, 171.

[DGCSAD11] Fernando De Goes, David Cohen-Steiner, Pierre Alliez, and Mathieu Desbrun, *An optimal transport approach to robust reconstruction and simplification of 2d shapes*, Computer Graphics Forum, vol. 30, Wiley, 2011, pp. 1593–1602.

[dGWH+15] Fernando de Goes, Corentin Wallez, Jin Huang, Dmitry Pavlov, and Mathieu Desbrun, *Power particles: an incompressible fluid solver based on power diagrams*, ACM Transactions on Graphics **34** (2015), no. 4, 50–1.

[Dob70] R. L. Dobrušin, *Definition of a system of random variables by means of conditional distributions* (Russian, with English summary), Teor. Verojatnost. i Primenen. **15** (1970), 469–497. MR0298716

[DR56] Jim Douglas Jr. and H. H. Rachford Jr., *On the numerical solution of heat conduction problems in two and three space variables*, Trans. Amer. Math. Soc. **82** (1956), 421–439, DOI 10.2307/1993056. MR84194

[ERSS17] Matthias Erbar, Martin Rumpf, Bernhard Schmitzer, and Stefan Simon, *Computation of optimal transport on discrete metric measure spaces*, Numer. Math. **144** (2020), no. 1, 157–200, DOI 10.1007/s00211-019-01077-z. MR4050090

[ES18] Montacer Essid and Justin Solomon, *Quadratically regularized optimal transport on graphs*, SIAM J. Sci. Comput. **40** (2018), no. 4, A1961–A1986, DOI 10.1137/17M1132665. MR3820384

[FCVP17] Jean Feydy, Benjamin Charlier, François-Xavier Vialard, and Gabriel Peyré, *Optimal transport for diffeomorphic registration*, MICCAI, 2017.

[FJF56] Lester Randolph Ford Jr. and Delbert Ray Fulkerson, *Solving the transportation problem*, Management Science **3** (1956), no. 1, 24–32.

[FO11] Brittany D. Froese and Adam M. Oberman, *Convergent finite difference solvers for viscosity solutions of the elliptic Monge-Ampère equation in dimensions two and higher*, SIAM J. Numer. Anal. **49** (2011), no. 4, 1692–1714, DOI 10.1137/100803092. MR2831067

[GM96] Wilfrid Gangbo and Robert J. McCann, *The geometry of optimal transportation*, Acta Math. **177** (1996), no. 2, 113–161, DOI 10.1007/BF02392620. MR1440931

[GMMD14] Fernando de Goes, Pooran Memari, Patrick Mullen, and Mathieu Desbrun, *Weighted triangulations for geometry processing*, ACM Transactions on Graphics (TOG) **33** (2014), no. 3, 28.

[Hir03] Anil Nirmal Hirani, *Discrete exterior calculus*, ProQuest LLC, Ann Arbor, MI, 2003. Thesis (Ph.D.)–California Institute of Technology. MR2704508

[Hit41] Frank L. Hitchcock, *The distribution of a product from several sources to numerous localities*, J. Math. Phys. Mass. Inst. Tech. **20** (1941), 224–230, DOI 10.1002/sapm1941201224. MR4469

[HZTA04] Steven Haker, Lei Zhu, Allen Tannenbaum, and Sigurd Angenent, *Optimal mass transport for registration and warping*, International Journal of Computer Vision **60** (2004), no. 3, 225–240.

[JKO98] Richard Jordan, David Kinderlehrer, and Felix Otto, *The variational formulation of the Fokker-Planck equation*, SIAM J. Math. Anal. **29** (1998), no. 1, 1–17, DOI 10.1137/S0036141096303359. MR1617171

[Jon01] Frank Jones, *Lebesgue integration on Euclidean space*, Jones and Bartlett Publishers, Boston, MA, 1993. MR1204268

[Kan42] L. Kantorovitch, *On the translocation of masses*, C. R. (Doklady) Acad. Sci. URSS (N.S.) **37** (1942), 199–201. MR0009619

[KL51] S. Kullback and R. A. Leibler, *On information and sufficiency*, Ann. Math. Statistics **22** (1951), 79–86, DOI 10.1214/aoms/1177729694. MR39968

[Kle67] Morton Klein, *A primal method for minimal cost flows with applications to the assignment and transportation problems*, Management Science **14** (1967), no. 3, 205–220.

[KM15] Jonathan Korman and Robert J. McCann, *Optimal transportation with capacity constraints*, Trans. Amer. Math. Soc. **367** (2015), no. 3, 1501–1521, DOI 10.1090/S0002-9947-2014-06032-7. MR3286490

[KMT16] Jun Kitagawa, Quentin Mérigot, and Boris Thibert, *Convergence of a Newton algorithm for semi-discrete optimal transport*, J. Eur. Math. Soc. (JEMS) **21** (2019), no. 9, 2603–2651, DOI 10.4171/JEMS/889. MR3985609

[Koo41] Tjalling C Koopmans, *Exchange ratios between cargoes on various routes.*

[KSKW15] Matt Kusner, Yu Sun, Nicholas Kolkin, and Kilian Weinberger, *From word embeddings to document distances*, International Conference on Machine Learning, 2015, pp. 957–966.

[Lav17] Hugo Lavenant, *Harmonic mappings valued in the Wasserstein space*, J. Funct. Anal. **277** (2019), no. 3, 688–785, DOI 10.1016/j.jfa.2019.05.003. MR3954993

[LB01] Elizaveta Levina and Peter Bickel, *The earth mover's distance is the Mallows distance: Some insights from statistics*, Proc. ICCV, vol. 2, IEEE, 2001, pp. 251–256.

[LCCS18] Hugo Lavenant, Sebastian Claici, Edward Chien, and Justin Solomon, *Dynamical optimal transport on discrete surfaces*, SIGGRAPH Asia 2018 Technical Papers, ACM, 2018, p. 250.

[Lév15] Bruno Lévy, *A numerical algorithm for L_2 semi-discrete optimal transport in 3D*, ESAIM Math. Model. Numer. Anal. **49** (2015), no. 6, 1693–1715, DOI 10.1051/m2an/2015055. MR3423272

[LHO10] B. Li, F. Habbal, and M. Ortiz, *Optimal transportation meshfree approximation schemes for fluid and plastic flows*, Internat. J. Numer. Methods Engrg. **83** (2010), no. 12, 1541–1579, DOI 10.1002/nme.2869. MR2722566

[LM79] P.-L. Lions and B. Mercier, *Splitting algorithms for the sum of two nonlinear operators*, SIAM J. Numer. Anal. **16** (1979), no. 6, 964–979, DOI 10.1137/0716071. MR551319

[Lot08] John Lott, *Some geometric calculations on Wasserstein space*, Comm. Math. Phys. **277** (2008), no. 2, 423–437, DOI 10.1007/s00220-007-0367-3. MR2358290

[LR05] Grégoire Loeper and Francesca Rapetti, *Numerical solution of the Monge-Ampère equation by a Newton's algorithm* (English, with English and French summaries), C. R. Math. Acad. Sci. Paris **340** (2005), no. 4, 319–324, DOI 10.1016/j.crma.2004.12.018. MR2121899

[LS18] Bruno Lévy and Erica Schwindt, *Notions of optimal transport theory and how to implement them on a computer*, Computers and Graphics **72** (2018), 135–148.

[Maa11] Jan Maas, *Gradient flows of the entropy for finite Markov chains*, J. Funct. Anal. **261** (2011), no. 8, 2250–2292, DOI 10.1016/j.jfa.2011.06.009. MR2824578

[McC94] Robert John McCann, *A convexity theory for interacting gases and equilibrium crystals*, ProQuest LLC, Ann Arbor, MI, 1994. Thesis (Ph.D.)–Princeton University. MR2691540

[McC97] Robert J. McCann, *A convexity principle for interacting gases*, Adv. Math. **128** (1997), no. 1, 153–179, DOI 10.1006/aima.1997.1634. MR1451422

[McC01] Robert J. McCann, *Polar factorization of maps on Riemannian manifolds*, Geom. Funct. Anal. **11** (2001), no. 3, 589–608, DOI 10.1007/PL00001679. MR1844080

[MCCD13] Tomas Mikolov, Kai Chen, Greg Corrado, and Jeffrey Dean, *Efficient estimation of word representations in vector space*, arXiv:1301.3781 (2013).

[MCSK+17] Manish Mandad, David Cohen-Steiner, Leif Kobbelt, Pierre Alliez, and Mathieu Desbrun, *Variance-minimizing transport plans for inter-surface mapping*, ACM Transactions on Graphics **36** (2017), 14.

[Mém09] Facundo Mémoli, *Spectral Gromov–Wasserstein distances for shape matching*, Proc. ICCV Workshops, IEEE, 2009, pp. 256–263.

[Mém11] Facundo Mémoli, *Gromov-Wasserstein distances and the metric approach to object matching*, Found. Comput. Math. **11** (2011), no. 4, 417–487, DOI 10.1007/s10208-011-9093-5. MR2811584

[Mém14] ———, *The Gromov–Wasserstein distance: A brief overview*, Axioms **3** (2014), no. 3, 335–341.

[Mér11] Quentin Mérigot, *A multiscale approach to optimal transport*, Computer Graphics Forum, vol. 30, Wiley, 2011, pp. 1583–1592.

[Mil07] Stacy Miller, *The problem of redistricting: the use of centroidal Voronoi diagrams to build unbiased congressional districts*, Senior project, Whitman College (2007).

[Mir15] Jean-Marie Mirebeau, *Numerical resolution of Euler equations, through semi-discrete optimal transport*, Journées Équations aux Dérivées Partielles (2015), 1–16.

[MM16] Quentin Mérigot and Jean-Marie Mirebeau, *Minimal geodesics along volume-preserving maps, through semidiscrete optimal transport*, SIAM J. Numer. Anal. **54** (2016), no. 6, 3465–3492, DOI 10.1137/15M1017235. MR3576572

[MMC16] Grégoire Montavon, Klaus-Robert Müller, and Marco Cuturi, *Wasserstein training of restricted Boltzmann machines*, Advances in Neural Information Processing Systems, 2016, pp. 3718–3726.

[MMP87] Joseph S. B. Mitchell, David M. Mount, and Christos H. Papadimitriou, *The discrete geodesic problem*, SIAM J. Comput. **16** (1987), no. 4, 647–668, DOI 10.1137/0216045. MR899694

[Mon81] Gaspard Monge, *Mémoire sur la théorie des déblais et des remblais*, Histoire de l'Académie Royale des Sciences de Paris (1781).

[NGT15] Lipeng Ning, Tryphon T. Georgiou, and Allen Tannenbaum, *On matrix-valued Monge-Kantorovich optimal mass transport*, IEEE Trans. Automat. Control **60** (2015), no. 2, 373–382, DOI 10.1109/TAC.2014.2350171. MR3310164

[Oli87] V. I. Oliker, *Near radially symmetric solutions of an inverse problem in geometric optics*, Inverse Problems **3** (1987), no. 4, 743–756. MR928051

[OP89] V. I. Oliker and L. D. Prussner, *On the numerical solution of the equation* $(\partial^2 z/\partial x^2)(\partial^2 z/\partial y^2) - ((\partial^2 z/\partial x \partial y))^2 = f$ *and its discretizations. I*, Numer. Math. **54** (1988), no. 3, 271–293, DOI 10.1007/BF01396762. MR971703

[Opt19] *OptimalDistricts.org*, 2019, Online (optimaldistricts.org); accessed August 8, 2019.

[Orl97] James B. Orlin, *A polynomial time primal network simplex algorithm for minimum cost flows*, Math. Programming **78** (1997), no. 2, Ser. B, 109–129, DOI 10.1016/S0025-5610(97)00011-7. Network optimization: algorithms and applications (San Miniato, 1993). MR1486304

[Ott01] Felix Otto, *The geometry of dissipative evolution equations: the porous medium equation.*

[Pas15] Brendan Pass, *Multi-marginal optimal transport: theory and applications*, ESAIM Math. Model. Numer. Anal. **49** (2015), no. 6, 1771–1790, DOI 10.1051/m2an/2015020. MR3423275

[PC19] Gabriel Peyré and Marco Cuturi, *Computational optimal transport*, Foundations and Trends in Machine Learning **11** (2019), no. 5-6, 355–607.

[PCS16] Gabriel Peyré, Marco Cuturi, and Justin Solomon, *Gromov–Wasserstein averaging of kernel and distance matrices*, International Conference on Machine Learning, 2016, pp. 2664–2672.

[PCVS17] Gabriel Peyré, Lénaïc Chizat, François-Xavier Vialard, and Justin Solomon, *Quantum entropic regularization of matrix-valued optimal transport*, European J. Appl. Math. **30** (2019), no. 6, 1079–1102, DOI 10.1017/s0956792517000274. MR4028471

[Pey10] Gabriel Peyré, *Optimal transport with Benamou–Brenier algorithm*, 2010.

[Pey15] Gabriel Peyré, *Entropic approximation of Wasserstein gradient flows*, SIAM J. Imaging Sci. **8** (2015), no. 4, 2323–2351, DOI 10.1137/15M1010087. MR3413589

[Pla12] A. Plakhov, *Billiards, scattering by rough obstacles, and optimal mass transportation*, J. Math. Sci. (N.Y.) **182** (2012), no. 2, 233–245, DOI 10.1007/s10958-012-0744-0. Translated from Sovrem. Mat. Prilozh., Vol. 71, 2011. MR3141344

[PP08] Kaare Brandt Petersen and Michael Syskind Pedersen, *The matrix cookbook*, Technical University of Denmark (2008).

[PPO14] Nicolas Papadakis, Gabriel Peyré, and Edouard Oudet, *Optimal transport with proximal splitting*, SIAM J. Imaging Sci. **7** (2014), no. 1, 212–238, DOI 10.1137/130920058. MR3158785

[PW09] Ofir Pele and Michael Werman, *Fast and robust earth mover's distances*, Proc. ICCV, IEEE, 2009, pp. 460–467.

[RMO93] Ravindra K. Ahuja, Thomas L. Magnanti, and James B. Orlin, *Network flows*, Prentice Hall, Inc., Englewood Cliffs, NJ, 1993. Theory, algorithms, and applications. MR1205775

[Roc15] Ralph Tyrell Rockafellar, *Convex analysis*, Princeton University Press, 2015.

[RTG00] Yossi Rubner, Carlo Tomasi, and Leonidas J Guibas, *The earth mover's distance as a metric for image retrieval*, International journal of computer vision **40** (2000), no. 2, 99–121.

[San15] Filippo Santambrogio, *Optimal transport for applied mathematicians*, Progress in Nonlinear Differential Equations and their Applications, vol. 87, Birkhäuser/Springer, Cham, 2015. Calculus of variations, PDEs, and modeling. MR3409718

[San17] Filippo Santambrogio, *{Euclidean, metric, and Wasserstein} gradient flows: an overview*, Bull. Math. Sci. **7** (2017), no. 1, 87–154, DOI 10.1007/s13373-017-0101-1. MR3625852

[SBD07] Lukas Svec, Sam Burden, and Aaron Dilley, *Applying Voronoi diagrams to the redistricting problem*, The UMAP Journal **28** (2007), no. 3, 313–329.

[SCSJ17] Matthew Staib, Sebastian Claici, Justin M Solomon, and Stefanie Jegelka, *Parallel streaming Wasserstein barycenters*, Advances in Neural Information Processing Systems, 2017, pp. 2644–2655.

[SDGP+15] Justin Solomon, Fernando De Goes, Gabriel Peyré, Marco Cuturi, Adrian Butscher, Andy Nguyen, Tao Du, and Leonidas Guibas, *Convolutional Wasserstein distances: Efficient optimal transportation on geometric domains*, ACM Transactions on Graphics (TOG) **34** (2015), no. 4.

[Set99] J. A. Sethian, *Fast marching methods*, SIAM Rev. **41** (1999), no. 2, 199–235, DOI 10.1137/S0036144598347059. MR1684542

[SG76] Sartaj Sahni and Teofilo Gonzalez, *P-complete approximation problems*, J. Assoc. Comput. Mach. **23** (1976), no. 3, 555–565, DOI 10.1145/321958.321975. MR408313

[SGB13] Justin Solomon, Leonidas Guibas, and Adrian Butscher, *Dirichlet energy for analysis and synthesis of soft maps*, Computer Graphics Forum, vol. 32, Wiley, 2013, pp. 197–206.

[She17] Jonah Sherman, *Generalized preconditioning and undirected minimum-cost flow*, Proceedings of the Twenty-Eighth Annual ACM-SIAM Symposium on Discrete Algorithms, SIAM, Philadelphia, PA, 2017, pp. 772–780, DOI 10.1137/1.9781611974782.49. MR3627778

[SK67] Richard Sinkhorn and Paul Knopp, *Concerning nonnegative matrices and doubly stochastic matrices*, Pacific J. Math. **21** (1967), 343–348. MR210731

[Sla50] Morton Slater, *Lagrange multipliers revisited*, Cowles Commission Discussion Paper (1950), no. 403, 1–13.

[Sol17] Justin Solomon, *Computational optimal transport*, Snapshots of Modern Mathematics from Oberwolfach (2017), no. 8, 1–15.

[SPKS16] Justin Solomon, Gabriel Peyré, Vladimir G Kim, and Suvrit Sra, *Entropic metric alignment for correspondence problems*, ACM Transactions on Graphics (TOG) **35** (2016), no. 4, 72.

[SRGB14a] Justin Solomon, Raif Rustamov, Leonidas Guibas, and Adrian Butscher, *Earth mover's distances on discrete surfaces*, ACM Transactions on Graphics (TOG) **33** (2014), no. 4, 67. DOI 10.1145/2601097.2601175

[SRGB14b] ______, *Wasserstein propagation for semi-supervised learning*, International Conference on Machine Learning, 2014, pp. 306–314.

[SRGB16] ______, *Continuous-flow graph transportation distances*, arXiv:1603.06927 (2016).

[STTP14] Yuliy Schwartzburg, Romain Testuz, Andrea Tagliasacchi, and Mark Pauly, *High-contrast computational caustic design*, ACM Transactions on Graphics (TOG) **33** (2014), no. 4, 74.

[Uza68] Hirofumi Uzawa, *Iterative methods for concave programming*, Studies in Linear and Non-Linear Programming **2** (1968), 154.

[Var67] S. R. S. Varadhan, *On the behavior of the fundamental solution of the heat equation with variable coefficients*, Comm. Pure Appl. Math. **20** (1967), 431–455, DOI 10.1002/cpa.3160200210. MR208191

[Vas69] Leonid Nisonovich Vaseršteĭn, *Markov processes over denumerable products of spaces, describing large systems of automata*, Problemy Peredachi Informatsii **5** (1969), no. 3, 64–72.

[Vil03] Cédric Villani, *Topics in optimal transportation*, Graduate Studies in Mathematics, vol. 58, American Mathematical Society, Providence, RI, 2003. MR1964483

[Vil08] ______, *Optimal transport: Old and new*, vol. 338, Springer, 2008.

[Wan96] Xu-Jia Wang, *On the design of a reflector antenna*, Inverse Problems **12** (1996), no. 3, 351–375, DOI 10.1088/0266-5611/12/3/013. MR1391544

[Wat81] D. F. Watson, *Computing the n-dimensional Delaunay tessellation with application to Voronoĭ polytopes*, Comput. J. **24** (1981), no. 2, 167–172, DOI 10.1093/comjnl/24.2.167. MR619577

Massachusetts Institute of Technology, 32 Vassar Street, Cambridge, Massachusetts 02139

Email address: `j.solomon@mit.edu`

SELECTED PUBLISHED TITLES IN THIS SERIES

77 **Pablo A. Parrilo and Rekha R. Thomas, Editors,** Sum of Squares: Theory and Applications, 2020

76 **Keenan Crane, Editor,** An Excursion Through Discrete Differential Geometry, 2020

75 **Michael Damron, Firas Rassoul-Agha, and Timo Seppäläinen, Editors,** Random Growth Models, 2018

74 **Jan Bouwe van den Berg and Jean-Philippe Lessard, Editors,** Rigorous Numerics in Dynamics, 2018

73 **Kasso A. Okoudjou, Editor,** Finite Frame Theory, 2016

72 **Van H. Vu, Editor,** Modern Aspects of Random Matrix Theory, 2014

71 **Samson Abramsky and Michael Mislove, Editors,** Mathematical Foundations of Information Flow, 2012

70 **Afra Zomorodian, Editor,** Advances in Applied and Computational Topology, 2012

69 **Karl Sigmund, Editor,** Evolutionary Game Dynamics, 2011

68 **Samuel J. Lomonaco, Jr., Editor,** Quantum Information Science and Its Contributions to Mathematics, 2010

67 **Eitan Tadmor, Jian-Guo Liu, and Athanasios E. Tzavaras, Editors,** Hyperbolic Problems: Theory, Numerics and Applications, 2009

66 **Dorothy Buck and Erica Flapan, Editors,** Applications of Knot Theory, 2009

65 **L. L. Bonilla, A. Carpio, J. M. Vega, and S. Venakides, Editors,** Recent Advances in Nonlinear Partial Differential Equations and Applications, 2007

64 **Reinhard C. Laubenbacher, Editor,** Modeling and Simulation of Biological Networks, 2007

63 **Gestur Ólafsson and Eric Todd Quinto, Editors,** The Radon Transform, Inverse Problems, and Tomography, 2006

62 **Paul Garrett and Daniel Lieman, Editors,** Public-Key Cryptography, 2005

61 **Serkan Hoşten, Jon Lee, and Rekha R. Thomas, Editors,** Trends in Optimization, 2004

60 **Susan G. Williams, Editor,** Symbolic Dynamics and its Applications, 2004

59 **James Sneyd, Editor,** An Introduction to Mathematical Modeling in Physiology, Cell Biology, and Immunology, 2002

58 **Samuel J. Lomonaco, Jr., Editor,** Quantum Computation, 2002

57 **David C. Heath and Glen Swindle, Editors,** Introduction to Mathematical Finance, 1999

56 **Jane Cronin and Robert E. O'Malley, Jr., Editors,** Analyzing Multiscale Phenomena Using Singular Perturbation Methods, 1999

55 **Frederick Hoffman, Editor,** Mathematical Aspects of Artificial Intelligence, 1998

54 **Renato Spigler and Stephanos Venakides, Editors,** Recent Advances in Partial Differential Equations, Venice 1996, 1998

53 **David A. Cox and Bernd Sturmfels, Editors,** Applications of Computational Algebraic Geometry, 1998

52 **V. Mandrekar and P. R.Masani, Editors,** Proceedings of the Norbert Wiener Centenary Congress, 1994, 1997

51 **Louis H. Kauffman, Editor,** The Interface of Knots and Physics, 1996

50 **Robert Calderbank, Editor,** Different Aspects of Coding Theory, 1995

49 **Robert L. Devaney, Editor,** Complex Dynamical Systems: The Mathematics Behind the Mandelbrot and Julia Sets, 1994

For a complete list of titles in this series, visit the
AMS Bookstore at **www.ams.org/bookstore/psapmseries/**.